MAKING TECHNOLOGY WORK

The successful application of new technology requires simultaneous consideration of the technical, economic, political, environmental, and international aspects of the innovation. This book presents fifteen cases of technology applications in the energy and environment sectors. The case studies include applications of solar, wind, fuel cell, nuclear, and coal combustion and emission control technologies. Both successes and failures are analyzed. The case studies demonstrate the importance of an interdisciplinary approach and of integrating technical and nontechnical aspects of the problem. The case studies are also used to introduce a toolbox of analytical techniques useful in the context of realistic technology application. These techniques include energy and mass balances, project financial analysis tools, the treatment of external costs and benefits, probabilistic risk assessment, learning curves, regression analysis, and life cycle costing. Each case study presents a description of the relevant technology at a level accessible to anyone familiar with elementary concepts in basic science and engineering. The book is addressed to upper-level undergraduate and graduate students in the natural sciences, engineering, and the social sciences who are interested in learning about problems of technology application, as well as technology practitioners in industry and government.

John M. Deutch is an institute professor at the Massachusetts Institute of Technology (MIT). He has been a member of the MIT faculty since 1970 and has served as chairman of the Department of Chemistry, Dean of Science, and Provost. Professor Deutch has published over 160 technical publications in physical chemistry, as well as numerous publications on technology, energy, international security, and public policy issues. Professor Deutch has also served as the Director of Energy Research (1977–1979), Acting Assistant Secretary for Energy Technology (1979), and Undersecretary (1979–1980) in the United States Department of Energy. He was Director of Central Intelligence from May 1995 to December 1996, and, from 1994 to 1995, served as Deputy Secretary of Defense and as Undersecretary of Defense for Acquisition and Technology from 1993–1994. Professor Deutch was a member of the President's Committee of Advisors on Science and Technology from 1997 to 2001 and is a director or trustee of the Council on Foreign Relations, Resources for the Future, and the Urban Institute.

Richard K. Lester is the founder and current director of the Industrial Performance Center and a professor of nuclear engineering at the Massachusetts Institute of Technology, where he has taught since 1979. His research focuses on productivity, product and process innovation, and the public management of technology. In recent years, Professor Lester has led a number of studies on national and regional productivity, industrial competitiveness, and innovation commissioned by governments and industrial groups around the world. His recent books include *The Productive Edge* (1998), an analysis of America's industrial resurgence during the 1990s, *Made by Hong Kong* (1997), co-authored with Suzanne Berger, and *Made in America: Regaining the Productive Edge* (1989), co-authored with Michael Dertouzos and Robert Solow. (With more than 300,000 copies in print in eight languages, *Made in America* is the best-selling title in the history of the MIT Press.) Professor Lester has also published widely in the fields of energy and the environment, and is co-author of *Radioactive Waste: Management and Regulation* (1978).

Making Technology Work

Applications in Energy
and the Environment

JOHN M. DEUTCH

Massachusetts Institute of Technology

RICHARD K. LESTER

Massachusetts Institute of Technology

CAMBRIDGE
UNIVERSITY PRESS

PUBLISHED BY THE PRESS SYNDICATE OF THE UNIVERSITY OF CAMBRIDGE
The Pitt Building, Trumpington Street, Cambridge, United Kingdom

CAMBRIDGE UNIVERSITY PRESS
The Edinburgh Building, Cambridge CB2 2RU, UK
40 West 20th Street, New York, NY 10011-4211, USA
477 Williamstown Road, Port Melbourne, VIC 3207, Australia
Ruiz de Alarcón 13, 28014 Madrid, Spain
Dock House, The Waterfront, Cape Town 8001, South Africa

http://www.cambridge.org

First published 2004

Printed in the United States of America

Typeface Minion 10.5/13.5 pt.   *System* LATEX 2_ε [TB]

A catalog record for this book is available from the British Library.

Library of Congress Cataloging in Publication Data
Deutch, John M., 1938–
Making technology work : strategies and techniques in energy and the
environment / John M. Deutch, Richard K. Lester.
p.   cm.
Includes bibliographical references and index.
ISBN 0-521-81857-5 – ISBN 0-521-52317-6 (pb.)
1. Power resources.   2. Environmental protection.   I. Lester, Richard K.
(Richard Keith), 1954–   II. Title.
TJ163.155.A1D48   2003
621.042 – dc21   2003046103

ISBN 0 521 81857 5 hardback
ISBN 0 521 52317 6 paperback

To A.E.C. and P.L.D.

Contents

Preface

This book grows out of a multidisciplinary course we have taught to MIT graduate and undergraduate students for over a decade. The course, "Application of Technology," is designed to introduce our students to the complex task of applying new technologies for economic, social or environmental purposes. Our goal in the course – and in this book – is to present insights, approaches, and analytical tools that are useful in such situations. This is an especially important subject for students educated in the sciences or engineering disciplines. Although most of these students will encounter complex problems of technology application in the course of their professional careers, their education today is focused on problems within their particular technical discipline. The solution of such problems may be crucial to technology invention and development but only a small part of what is required for successful technology application.

A central theme of this book is that students in the sciences and engineering should recognize the importance of moving away from thinking solely about the creation of new technology to thinking also about its responsible and effective application. Finding the right balance between these two ways of thinking is a fundamental challenge for technology practitioners – and for educators in science and engineering as well.

A second and related challenge is to move from working within the boundaries of a single discipline to integrating across disciplines. The responsible and effective application of most technologies requires a combination of scientific, engineering, economic, manufacturing, organizational, legal, political, and, increasingly, international considerations. And the prospects for successful application will be greater to the degree that these issues are addressed simultaneously and comprehensively from the earliest stages of technology development. People from many disciplinary backgrounds need to be involved in the process, and this creates the third challenge for the technology practitioner: the need to shift from working only with like-minded colleagues to working with people who have very different ways of defining and solving problems. In the past, scientists and engineers have typically left problems of technology application to nontechnical professionals – financiers, marketing specialists, lawyers, accountants, and others.

The examples of technology applications in this book are drawn mainly from the energy and environmental fields. But the tools, techniques, and approaches will be useful in a wide range of other fields too, such as information and communications, medical technology, and national security.

In each of the case studies we present here – which include applications of wind, solar, fuel cell, nuclear, fossil, and energy conservation technologies – a description of the relevant technology is provided at a level accessible to anyone who is familiar with elementary concepts in basic science and engineering. Thus, although the book is targeted at advanced undergraduate and graduate students in the sciences and engineering, readers with a wide range of backgrounds will benefit from the material, including technology practitioners and participants in mid-career and executive education programs.

We are deeply indebted to the Alfred P. Sloan Foundation for its support of this project. We are also grateful for the financial assistance provided by the Office of the Dean of Engineering at MIT. We owe much to our "Application of Technology" teaching assistants during the preparation of this book, especially Eugene Bae and David Ward, and to Anita Kafka for her help, especially with the illustrations. Finally, we wish to acknowledge the support and encouragement provided by our editor at Cambridge University Press, Scott Parris.

1

Introduction

Applying new technology in our society is invariably a challenge, and those who try do not always succeed. New technologies are frequently of large scale, involve significant environmental or social consequences and must adhere to a complex framework of governmental rules and regulations whose economic impact may be far-reaching. Issues such as opposition to nuclear power, concern over the environmental effects of burning coal, the ethical dilemmas of stem cell research, and the threats to privacy, intellectual property, and even national security associated with the growing use of the Internet fill the daily newspapers. Learning how to manage the often-competing interests that come into play when new technologies are deployed in society will be increasingly important, especially for scientists and engineers whose professional lives are dedicated to the task of harnessing technology for economic and social ends.

Today the education of scientists and engineers in U.S. universities is still strongly influenced by the conventional view of technological innovation as a linear process. In this view, innovation proceeds through distinct stages: (1) research – the first step of knowledge creation, usually by scientists in a laboratory; (2) development – the step of reducing the knowledge to practice, normally the responsibility of the engineer; and (3) application – the crucial step of implementing a technology, mainly the province of nontechnical professionals, such as managers, financiers, lawyers, politicians, or public-interest advocates. Scientific and engineering education is organized according to this linear perspective. The curriculum of a typical student in physics, chemistry, and chemical or electrical engineering understandably stresses depth in the discipline and research skills. The student's experience, however, includes little if any exposure to other disciplines, to techniques that are useful for analyzing the multidimensional aspects of technology application, or to working with a multidisciplinary group to address a complex technology problem. Yet most science and engineering students will encounter the broad range of problems associated with technology application almost immediately in their professional careers. They will be relatively unprepared to deal with these problems.

As we shall see repeatedly throughout this book, successful application of technology requires that simultaneous consideration be given to the technical and nontechnical aspects of the situation, because of the interrelationships among these elements. This perspective on technology innovation stresses integration and differs significantly from the traditional linear view. A necessary consequence is that the application of technology cannot be left to nontechnical professionals alone. Scientists and engineers must be actively involved not only in creating new technology options but also in the complex process of determining the circumstances of technology application. The challenge for practicing technologists – and for education in science and engineering – is to achieve a better balance between inventing new technology and responsible application.

This book is addressed to science and engineering students (both graduate and undergraduate) who are aware of the limitations of the current educational approach and who are interested in learning about problems of technology application. The book is an outgrowth of the multidisciplinary subject called "Application of Technology – Case Studies in Energy and the Environment," developed with support from the Alfred P. Sloan Foundation and taught at MIT since 1992. The case studies presented in the book include applications of nuclear, coal-burning, solar, wind, and energy conservation technologies. Each case study presents a description of the relevant technology at a level accessible to anyone who is familiar with elementary concepts in basic science and engineering.

Each case study integrates technical analysis with the economic, political, environmental, and social aspects of the technology application under consideration. Where appropriate, international considerations are also included. It is the integration of these aspects that both defines the barriers to technology application and points to possible solutions. To take just one example, it is often said that nuclear waste disposal is a political rather than a technical problem. This distinction implies that the implementation of this technology is the responsibility of politicians rather than scientists and engineers, and that the political constraints are separable from technical considerations of repository design, siting, construction, operation, and cost. But this kind of separation is impossible. As we shall see, responsible progress requires the simultaneous consideration of political, economic, environmental, and technical factors.

Of course, not all new technologies present the same range of issues as energy technologies. Information technologies such as computers, telecommunication networks, and the Internet clearly do not raise the sort of thorny environmental issues that figure in the application of many energy systems. On the other hand, other public policy issues are critical in the information and communications industries. For example, there are no internationally accepted rules for encryption of personal or commercial communications; fundamental privacy issues are raised by the Internet and electronic commerce; and the relative economics of long-distance land-line, wireless, satellite, and cable transmission technologies are importantly affected by government policies regarding taxes, antitrust, and price regulation. The central point is that the successful application of technology in any industry – from health care to transportation, from energy to biotechnology – requires simultaneous consideration of technical and nontechnical factors. Although the case studies in this book are drawn from the energy and environmental sectors, the reader should gain an appreciation for the integrated approach required for progress on a wide range of applications of complex technologies.

Serious students are not interested in merely hearing "war stories" about the difficulties of applying technology. Rather, they wish to learn techniques and skills needed to address the problems they will encounter in their professional careers. The second objective of the case studies is thus to present techniques in the context of realistic application that the student will be able to apply in new situations. The reader should expect to accumulate a "toolbox" of techniques that will be useful in analyzing new problems. Examples of the types of tools that are introduced in this book include: (1) energy and materials balances; (2) cost-benefit analysis; (3) the treatment of external costs and benefits; (4) present worth analysis of costs and benefits; (5) probabilistic risk assessment; and (6) life cycle costing.

Every public policy issue involves many groups that have an interest in the outcome – entrepreneurs, politicians, financiers, public interest groups, and others. By understanding the consequences of technology applications for each group, the different stakeholders can expect to achieve their objectives more readily and more responsibly. In most cases, of course, there is no perfect outcome. There are invariably winners and losers, and some interests that come closer to being satisfied than others. No single outcome can be identified as – or is perceived to represent – "the public interest." Still, understanding the views of all the stakeholders in the decision-making process helps to reach the most satisfactory resolution.

In the following paragraphs, we illustrate the approach that will be taken in subsequent chapters with two brief examples.

PAPER OR FOAM PLASTIC CUPS?

In December 1990, in a highly publicized press release, the McDonald's Corporation and the Environmental Defense Fund (EDF), a nationally known environmental public interest group, jointly announced that McDonald's had decided to replace polystyrene foam food and drink containers with paper containers to achieve environmental benefit. Whatever they are made of, whether plastic or paper, these boxes and cups are waste byproducts of producing and consuming McDonald's meals. At first glance, the decision appeared uncontroversial – paper is recyclable and biodegradable, whereas plastic is not – and McDonald's, a good corporate citizen, decided to switch from one product to another because a public interest group pointed out the environmental benefits of doing so. (The decision was also economically rational. The switch did not actually involve any extra costs to McDonald's, and the favorable public response was expected to yield economic benefits to the corporation.)

Then Martin Hocking, a professor of chemistry at the University of Victoria in British Columbia, published an article questioning whether the selection of paper over plastic did, in fact, have environmental merit. Hocking compared the waste streams generated in the production of paper and plastic cups.[1] Although Hocking's analysis has been criticized by paper advocates,[2] it is instructive on two points that often arise in technology applications.

---

[1] Martin B. Hocking, *Science* 251, 504–5 (1991).
[2] Red Caveney, *Science* 252, 1362 (1991); Henry Wells, *Science* 252, 1361 (1991).

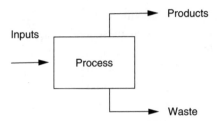

Figure 1.1. Conceptual diagram of an industrial process.

Internalizing External Costs

The conceptual flow diagram in Figure 1.1 applies to any industrial process. The process takes inputs (raw materials) and converts them to products, while generating wastes that are assumed to be valueless. The wastes may be in solid, liquid, or gaseous form, and their disposal can impose a burden on the environment, for example, by requiring unsightly landfills or by polluting streams or the atmosphere.

In the past, the environmental impacts of waste streams were usually disregarded, and accordingly, the costs of those impacts that were actually borne by the producer were negligible. They were "externalized." The cost to the producer of manufacturing the product was simply the cost of the inputs and the cost of building and operating the production facility.

Today a great deal of attention is given to reducing the environmental burdens of waste disposal, and accordingly, the monetary costs of disposing of wastes have risen. Companies can be expected to act rationally in selecting processes that minimize the overall cost of production, including the cost of waste disposal. In some cases the monetary cost of disposal borne by the company may also include the residual cost to the environment inflicted by the waste streams, that is, the residual environmental costs are "internalized." If the cost to the company for waste disposal properly reflects the environmental burden, then one can expect that the company will select the process that is most efficient for society, which is the process that minimizes the total social cost. However, even under these circumstances, environmental groups may not be satisfied. First, there may be disagreement as to whether the residual environmental costs have been correctly internalized. Second, there will always be some environmental advocates who place greater value on reducing harmful environmental impacts than can be justified on the basis of economic optimization. One cannot expect to satisfy all environmental concerns, any more than one should expect to satisfy all the concerns of any interest group. However, the governing principle is clear: To the extent possible, environmental costs to society should be internalized, in the sense that these costs should be included in the total cost incurred by the company in producing its product, and therefore in the price paid for the product by the company's customers.

The problem arises when the cost of waste disposal paid by the company does not reflect the actual environmental burden. This is surely the case for McDonald's, where carryout food is sold in paper or plastic containers. When the consumer throws the wrappers away, someone else pays for the cost of disposal. The environmental costs are "externalized," in the sense that they are not included in the price the company

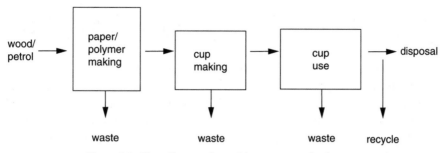

Figure 1.2. Flow diagram for making paper or plastic cups.

pays for waste disposal. Under these circumstances, the public is justified in insisting on regulations that force the company to make the choices it would have made had the external costs been internalized. Public interest groups deserve support in urging companies to take actions that are consistent with taking external costs into account in their business decisions.

The controversy here is essentially about the external environmental costs associated with the use of paper versus plastic. Because the environmental impacts occur outside any market framework, it is necessary to estimate the magnitude of the environmental burden and its associated costs. Different stakeholders (such as business firms and environmentalists) will have different perceptions of the severity of these environmental impacts. Resolving disputes about the magnitude of the external costs and how they should be taken into account is one of the central obstacles to the application of many technologies.

What is the System?

Hocking's analysis is interesting because he examines several intermediate steps in the overall process of making and using paper and plastic cups. Hocking includes in his analysis the paper-making process, the process of making polystyrene polymer from petroleum, and the potential for recycling used cups of both types.

A simplified diagram of the cup-making process considered by Hocking is given in Figure 1.2. With the intermediate steps of paper and plastic manufacturing included, a simplified summary of Hocking's analysis is given below:

Category	Foam cup	Paper cup
input: wood	0 g	33 g
input: petroleum	3.2 g	4.1 g
weight	1.5 g	10.1 g
cost	x	$2.5x$
recycle	some	low
biodegrades	no	yes
burns	clean	clean

By including the step of paper-making in the process, Hocking reverses the traditional conclusion that paper cups use less petroleum than plastic. The key to his analysis and the subsequent debate over the relative environmental benefits of paper versus plastic is the definition of the process system under consideration. Different conclusions are reached depending upon how the system boundary is drawn.

Critics from the paper industry object to Hocking's estimate of the amount of petroleum needed to make a paper cup (they estimate less than 2 g compared with his estimate of 4.1 g), and they take a different view of both the volume needed for landfill disposal and the potential for recycling paper versus plastic. The balance is not clear. What do you think?

ELECTRIC VEHICLES IN CALIFORNIA

In 1990, the California Air Resources Board (CARB), a department of the California Environmental Protection Agency, mandated that by 1998 2% of all cars and light trucks sold in California by the major automobile manufacturers must be "zero emission vehicles" (ZEV). The target rose to 10% of new vehicle sales by the year 2003 and increased further thereafter. Other states followed California's lead.

In taking this action the California government was seeking to compensate for an external environmental cost – the effect of auto emissions on air quality, especially in urban areas in southern California – that was not being adequately taken into account by the market. By mandating the dates by which fixed percentages of zero emission vehicles would have to be introduced, the state was pursuing a "command and control" approach to internalizing these costs.

An alternative regulatory mechanism is taxation. If the state taxes polluting vehicles in proportion to the amount of pollution they emit into the atmosphere, there will be an economic incentive for automobile manufacturers to introduce lower emitting vehicles. Presumably there is some level of taxation that would result in the same improvement in atmospheric air quality as the command and control approach. And there is merit in relying on an indirect taxation mechanism rather than the direct regulatory approach, because the former permits private companies to respond in a manner that is most efficient for them instead of being required to conform to a single design solution. (As it turns out, CARB later replaced its original 1990 requirements with more flexible targets, although key aspects of the command and control approach were retained.)

At the time of the original regulations – and still today – only electric vehicles conform to the zero emission standard. At the present level of battery technology, electric vehicles still have severe performance constraints, including limitations on acceleration, battery recharging time, and range. In order to achieve reasonable round trip travel ranges of about 50 mi, the electric vehicles must be quite heavy (due to the weight of the batteries required) and are relatively costly. The electric vehicles cannot travel very far from an electrical recharging point.

The appropriate environmental objective is to improve air quality by reducing emissions below current levels. The ultimate objective is not to introduce electric vehicles; that is merely the means to the end, and there are several other alternatives to the current conventional gasoline-powered car that should also be considered, for example:

Base case	Alternatives
Conventional gasoline-powered vehicles	Battery electric vehicles
	New low or ultra-low emitting gasoline-powered vehicles
	Compressed natural gas-powered vehicles
	Hybrid electric vehicles

A comparative analysis of these alternatives is not presented here. Our more limited purpose is to point out that by relying on regulations that effectively specify a particular vehicle type – ZEV – consideration of other interesting alternatives is precluded. For example, the hybrid electric vehicle (HEV) carries on board a small constant rpm engine that can very efficiently and with very low emissions charge batteries. The fuel for this small engine generator could be either natural gas or gasoline. With this design, the HEV circumvents the major disadvantage of the pure electric vehicle, because the small electric generator permits long-distance trips. Emissions per mile traveled are dramatically reduced compared with conventional vehicles; this is due to the very low gasoline consumption that is achieved by the HEV as a result of relying on the constant rpm engine. Specifying a particular system in the regulations may not lead to the desired outcome.

Moreover, in this case (as in the previous example of paper versus plastic) the issue arises as to what system is under consideration? If the atmospheric emissions from a pure electric vehicle are compared with those from a gasoline vehicle, it is clear that the pure electric vehicle has the lower emissions. But if a comparison is made between the system comprising ZEV and its attributable utility generation and the gasoline-powered car and its fuel supply system, the outcome for air quality is less clear. The result will partly depend on whether the electricity is generated by nuclear, coal, or oil-fired power plants. It is always important to define the system under consideration in comparative analysis.

For the comparison between conventional gasoline-powered autos and electric vehicles, consider the following, simpler question: Which is more energy efficient? Suppose that all the electricity generated comes from oil (which is actually not true in California.) The comparison is shown in Figure 1.3. For the case of the electric vehicle, there is an efficiency loss of two thirds associated with the conversion from oil to electricity, and a further 25% loss incurred in transmitting the electricity from the power plant to the wall plug used to charge the electric vehicle. If it takes 1 kwhr of electric energy to drive the electric vehicle 1 mi, we find that 13,650 British Thermal Units (BTUs) of oil are required for 1 mi of travel in this case.[3] For the conventional gasoline-powered car, there is a 10% loss associated with the conversion of oil to gasoline at the refinery. If the

---

[3] A British Thermal Unit (BTU) is the amount of energy required to increase the temperature of a cubic foot of water by 1°F. The energy industry in the United States has unfortunately not yet adopted metric (SI) units. 1 kwhr of energy is equivalent to 3,412 BTU.

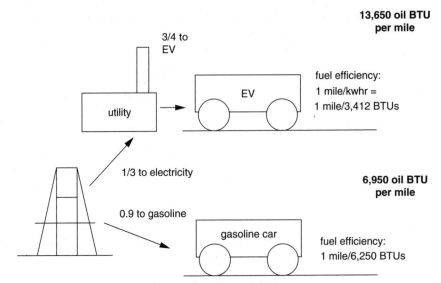

Figure 1.3. Oil required for an electric vehicle (EV) to travel one mile compared with a gasoline-powered car (assumed car mileage of 20 mpg of gasoline @ 125,000 BTU per gal).

car does 20 mpg of gasoline, this translates into an oil requirement of approximately 7,000 BTUs of oil per mile. Thus on energy efficiency grounds, given today's technology, a conventional gasoline-powered car is almost twice as efficient as an electric vehicle charging from an oil-fired electric power plant. The question is whether the emission advantages of the electric vehicle would override the considerable economic penalty revealed by this difference in energy efficiency.

The key insights from these brief examples are developed more fully in subsequent chapters. In the next chapter, on the production of gasohol fuel from corn, we further demonstrate the importance of clearly defining the boundaries of the system being analyzed. And in several later chapters we revisit the problem of external costs and consider how alternative ways of dealing with these costs can affect the application of new technology.

2

Gasohol

This chapter considers the question: Should national energy policy encourage the growing of corn to produce gasohol? Gasohol is the product of the conversion of corn or sugar to ethanol (ethyl alcohol), which is employed as a gasoline additive. Ordinarily, the term gasohol refers to a mixture of 10% ethanol and 90% gasoline. The idea of gasohol in current U.S. policy is simple: Use ethanol from corn to displace a portion of the gasoline for motor vehicles, thereby substituting a renewable energy source (corn) for a depletable energy source (petroleum). Because the purpose of government support for gasohol is to substitute for petroleum, we focus on the petroleum fuel and undertake a careful energy balance comparing the petroleum needed to produce gasohol with the petroleum that the gasohol displaces. The point is to identify the net petroleum displaced by gasohol.

TECHNICAL ASPECTS OF GASOHOL

The production of ethanol from corn requires several steps. First, the corn must be grown; then the starch (long-chain macromolecules made up of six-carbon sugars) must be separated from the corn. The starch is then hydrolyzed to glucose, which in turn is fermented to form ethanol. Finally, the ethanol is separated from the fermentation liquor by distillation.

Each of these steps requires energy. Much of the energy required for growing the corn comes from sunlight. But the intensive form of agriculture practiced in the United States, which results in high crop yields, requires a considerable amount of expensive premium fuels for fertilizers, farming, and harvesting. The premium fuels – natural gas, diesel fuel, oil, or gasoline – are directly usable either as boiler or transportation fuels, and are equivalent to the gasoline that the gasohol is intended to displace. In the case of gasoline and diesel fuel oil, moreover, a significant amount of energy is needed to refine the crude oil from which they are obtained. Similarly, a good deal of energy is required for the gasohol conversion processes of hydrolysis, fermentation, and distillation. In principle, the heat energy required for conversion can come from any source, including coal, but in practice either diesel or natural gas is most likely to be used in the United States.

In sum, both the production of corn and its conversion to ethanol require considerable amounts of premium fuel. So our analysis of the desirability of gasohol should focus on the net amount of premium fuel saved – the amount of gasoline displaced in motor vehicles minus the amount of premium fuel employed in producing and converting corn to ethanol. The actual amounts of premium energy required vary across locations

9

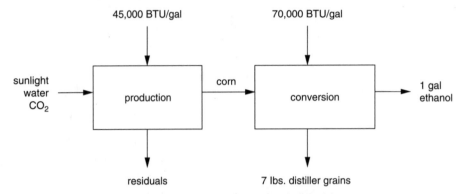

Figure 2.1. Flow diagram for production of ethanol from corn.

and fermentation facilities. A reasonable estimate of the premium energy required to grow and harvest the 0.4 bushels of corn (weighing approximately 20 lbs.) needed to produce 1 gal of ethanol is 45,000 BTU per gal. The fuel requirement for conversion to ethanol is in the range of 70,000 BTU per gal. It should be noted that estimates of the premium energy requirement for each of these steps vary widely. The variation is partly the result of actual differences in practice, and partly due to differences in the perspectives of the analysts.

The process of converting corn to ethanol produces an important by-product – the solids left in the fermentation liquor. These solids, called distillers dry grain, can be used as animal feed. Approximately 7 lbs. of distillers dry grain are produced for each gallon of ethanol. The flow diagram for producing ethanol from corn is presented in Figure 2.1.[1]

Energy Content of Ethanol and Gasoline

Gasoline is composed of a mixture of hydrocarbons. On a weight basis, ethanol has a lower heat of combustion than the hydrocarbon fuel it replaces because ethanol has a water molecule built in:

$$CH_3CH_2OH + 3O_2 \rightarrow 2CO_2 + 3H_2O.$$

The heat content of gasoline is 125,000 BTU per gal, while the heat content of ethanol is 84,000 BTU per gal. Thus, on an energy basis, 1 gal of ethanol is roughly equivalent to 0.67 gal of gasoline. On the other hand, ethanol does have some value as an octane booster – its research octane number (RON) is 95 as compared to 85 for unleaded gasoline.[2]

---

[1] Ethanol can also be produced from petrochemicals, and indeed approximately half the ethanol consumed annually comes from petroleum sources. Because gasohol is intended to displace petroleum products, this synthetic ethanol is not of direct concern. Synthetic ethanol does, however, influence the market price of ethanol.

[2] The octane number of gasoline is a figure of merit representing its resistance to premature detonation when exposed to heat and pressure in the combustion chamber of an internal-combustion engine, leading to engine "knock" and inefficient fuel use. The octane number of a sample of fuel is determined by burning

How Much Premium Fuel Will Gasohol Save?

The answer depends upon the assumptions underlying an energy balance calculation for the premium fuel involved in producing the ethanol. We consider several cases in turn:

a. *Energy Equivalence.* The premium energy required for ethanol production and conversion is the energy input, $45{,}000 + 70{,}000 = 115{,}000$ BTU per gal. The ethanol replaces gasoline based on its energy equivalence. If no by-product credits are included, this leads to a net loss of premium energy:

$$\text{Premium energy gain} = 84{,}000 - 115{,}000 = -31{,}000 \text{ BTU per gal.}$$

b. *By-product Credit.* The distiller dry grain can be taken as a credit. Because it is as useful for animal feed as corn, the 7 lbs. of distillers dry grain is equivalent to about one-third of the weight of the 0.4 bushels of corn required to produce 1 gal of ethanol. Thus, the by-product credit reduces the effective energy input for production by $(1/3) \times 45{,}000 = 15{,}000$ BTU per gal. This still leaves a small net loss of premium energy:

$$\text{Premium energy gain} = 84{,}000 - 100{,}000 = -16{,}000 \text{ BTU per gal.}$$

c. *Nonpremium Fuel Fired Conversion with By-product Credit.* If agricultural residues or coal are used instead of a premium fuel to provide the heat for ethanol conversion, and the by-product credit for distillers dry grain is also taken, there is a net gain in premium energy:

$$\text{Premium energy gain} = 84{,}000 - 30{,}000 = +54{,}000 \text{ BTU per gal.}$$

d. *Volume Equivalence.* Some assume that ethanol acts as a mileage extender. Its higher octane rating leads to better combustion characteristics for the entire gasohol mixture, and the ethanol is assumed to be equivalent to gasoline on a volume basis. With coal used for corn conversion and by-product credit taken, this leads to a large net gain in premium energy:

$$\text{Energy gain} = 125{,}000 - 30{,}000 = 95{,}000 \text{ BTU per gal.}$$

Each of these calculations involves an energy balance of premium fuels based on different assumptions about energy requirements. Yet, if the purpose of a gasohol program is to displace premium fuels, there must be agreement on the energy balance to be used. Most knowledgeable observers believe case "b" to be most reasonable, absent a large

---

the gasoline in an engine under controlled conditions, such as, spark timing, compression, engine speed, and load, until a standard level of knock occurs. Research octane number (RON) refers to one set of specified engine conditions.

and expensive effort to convert existing fermentation and distillation capacity to coal.[3] According to these assumptions, a corn-based gasohol program not only would not displace premium fuels, but would actually be a net consumer of about 16,000 BTUs of premium fuel per gal of ethanol produced.

ECONOMIC ASPECTS OF GASOHOL

If the purpose of gasohol is to displace gasoline, but gasohol requires more premium fuel than it displaces, why isn't the price of gasohol higher than that of gasoline, allowing market forces to take care of this inefficient allocation? The reason is that distorting tax credits have been granted to gasohol. The granting of these credits is what makes gasohol a national energy policy issue. If government assistance in the form of tax credits had not been sought or deemed necessary, advocates of gasohol would be free to compete in the market place without the need to convince anyone that a net positive amount of premium fuel would be displaced. But once an advocate seeks federal assistance, the validity of the claim must be examined.

In 1979, gasohol was given an exemption of \$.04 per gal from the then prevailing federal tax on gasoline. This is equivalent to a \$.40 per gal tax credit for ethanol in a 90/10 gasohol mixture. Because there are approximately 40 gal per barrel, this federal tax credit corresponds to a \$16 per barrel subsidy for ethanol production.

In addition, many states, particularly corn-growing states, waived state gasoline taxes on gasohol. For a state gasoline tax of \$.10 per gal, this is equivalent to an additional subsidy for ethanol production of \$1 per gal. In these states, therefore, ethanol is eligible for a total tax credit of \$1.40 per gal. The cost of producing ethanol from corn is about \$1.50 per gal. This cost includes the price of the premium fuel that is used in the production process. Because the refinery acquisition cost of gasoline is about \$.90 per gal, it is clear that ethanol would never displace gasoline if there was no subsidy. But the effect of the tax credit is to give gasohol a net cost advantage over gasoline of \$.08 per gal:

$$\text{cost of gasohol (\$/gal)} = [0.9 \, (\text{gal of gasoline}) \times 0.90 \, (\text{\$/gal})] \\ + [0.1 \, (\text{gal of ethanol}) \times 1.50 \, (\text{\$/gal})] \\ - 0.14 \, (\text{\$/gal tax credit}) \\ = 0.82 \, \text{\$/gal}$$

It is therefore not surprising that gasohol is being sold in those states that offer a tax credit.

From the national perspective, each gallon of ethanol is displacing, at best, 1/2 gal of premium fuel [see case c above]. This means that U.S. taxpayers are paying \$2.80 per gal of oil displaced or approximately \$112 per barrel. If the national objective is

---

[3] See, for example: Thomas R. Stauffer, "Gasohol: The costly road to autarky," John F. Kennedy School of Government discussion paper, May 1981.

to displace oil (either because it is a depletable resource or because we wish to reduce our dependence on imported oil), many alternatives are cheaper than gasohol, such as energy conservation.[4]

HOW FAR COULD GASOHOL USAGE BE PUSHED AS NATIONAL ENERGY POLICY?

In 2001, consumption of motor gasoline in the United States was about 8.5 million barrels per day (130 billion gal per year.) If gasohol made a 10% penetration into the gasoline market, this would require 85,000 barrels per day of ethanol production. This in turn would require the production of 520 million bushels of corn per year – about 5% of current corn production, requiring, at present yields, about 3.8 million acres under cultivation. A gasohol program of this magnitude would displace about 43,000 barrels per day of oil. This can be compared with the average daily consumption of petroleum in the United States of about 20 million barrels per day. Thus, even at the ambitious level of 10% penetration, gasohol would not significantly reduce our dependence on petroleum. At higher levels of penetration, ethanol derived from biomass would come into competition with corn for use as food – a significant long term issue.

This analysis is based on the use of corn as the raw material. Ethanol can also be made from petroleum, but then its use in gasoline is of interest only as an octane enhancer. In this application, it competes against other oxygenated fuel additives such as MTBE (methyltertiarybutylether).[5] On the other hand, if it were possible to make ethanol easily from the cellulose in wood, which is a plentiful and much cheaper biomass feedstock, the prospects for gasohol would be quite different. We discuss this possibility later in the chapter.

POLITICS OF GASOHOL

Gasohol production was pushed in the late seventies by Democratic senators from farm states – including Illinois, Iowa, Indiana, and South and North Dakota – reflecting the views of their constituents. The advocacy of gasohol is an excellent example of the political difficulties of adopting rational energy measures in a democratic society.

When federal tax credits for gasohol were first proposed, Edward Frieman, then the Director of Energy Research of the U.S. Department of Energy (DOE), was requested by Charles Duncan, the Secretary of Energy, to undertake a technical study of the prospects for gasohol.[6] A study group of experts chaired by Professor David Pimentel of Cornell

[4] Advances in the technology or energy efficiency of farming or corn conversion could reverse this conclusion. Also, important variations from the average premium energy inputs may occur from one farm or fermentation plant to another, making gasohol more attractive in some locations.
[5] MTBE [$CH_3-O-C(CH_3)_3$] is made from petroleum feed stock by the addition of methanol, CH_3OH, to isobutylene, $CH_2=C(CH_3)_2$. Since 1999, MBTE has been banned as an additive to gasoline because it has been found in ground water to which it gives a very unpleasant odor.
[6] At the time, one of the authors (Deutch) was Undersecretary of Energy at the DOE.

University was formed as a task force of the Energy Research Advisory Board (ERAB) to address the issue.

The ERAB gasohol report, issued in April 1980,[7] was immediately attacked by senators and congressmen from farm states, who criticized the task force conclusions and the composition of its membership as lacking in balance and objectivity. Congress called for a report by the General Accounting Office (GAO), which duly issued one criticizing the balance of views within the task force and some of its procedures.[8] However, the GAO did not find any substantive faults in either the task force's analysis or its conclusions, other than noting that it did not adopt an optimistic perspective.

The response of an important political interest group (in this case, farmers) to a technical/economic analysis that reaches conclusions which go against its interest is usually emotional and not substantive. The criticism of the ERAB gasohol report by both Congressional Democrats and the Executive Branch was driven by the fact that the technical analysis, however valid, had reached politically unpalatable conclusions. Those who undertake technical/economic analyses must recognize this risk and act accordingly. In general, the analysis should seek to define compromise outcomes that are "second best," but give opponents some room. It is also very important to forewarn both supporters and opponents about the likely conclusions of a potentially controversial study. Not doing so makes opponents all the more angry while providing insufficient time for supporters to marshal help. However, as the ERAB gasohol saga illustrates, it is not always possible to avoid the storm.

Today there is more political interest in protecting the environment than in reducing dependence on petroleum. Accordingly, a broad-based "alcohol fuels" lobby has formed that combines the interests of producers of methanol [CH_3OH], natural gas, and other oxygenated additives such as MTBE, as well as farmers, in order to promote the benefits of alcohol fuels for improving air quality. This coalition succeeded in incorporating provisions for alcohol fuels in the Clean Air Act of 1990. The Tax Reform Act of 1996 maintained the federal gasohol tax subsidy so that today there is a 4.4¢ per gal federal tax credit for gasohol out of a total federal excise tax of 18.3¢ per gal.

GASOHOL ELSEWHERE IN THE WORLD

Other nations have experience with ethanol and methanol as transportation fuels. Brazil, in particular, has adopted an ambitious program of pure ethanol vehicles with ethanol produced from sugar, not corn. The Brazilian situation is quite different from that in the United States. Despite the higher price of sugar compared to corn, the marginal cost of Brazilian sugar production is low, and the sugar cane by-product can also be used as boiler fuel. Moreover, the premium energy input for sugar production in Brazil is quite low, and less energy is required to convert sugar to ethanol than is required for the

---

[7] Report of the Gasohol Study Group, prepared for the Energy Research Advisory Board, U.S. Department of Energy, 1980.

[8] Report of the Comptroller General of the General Accounting Office, "Conduct of DOE's Gasohol Study Group: Issues and Observations," EMD 80–128, September 30, 1980.

conversion of corn. Thus, the analysis for ethanol in Brazil (or in other nations) is quite different from the one presented above for the United States. However, the Brazilian experience with ethanol as a transportation fuel has not been successful either, partly because petroleum prices have fallen in Brazil, sugar prices have strengthened, and the costs of distributing ethanol are larger than expected.

ETHANOL FROM CELLULOSE

The difficulties of producing ethanol from corn do not necessarily apply to other starting materials. Cellulose, along with lignin, forms the basis of the nonstarch component of corn, wood, and other plants. Cellulose is composed of five- and six-carbon sugar chains, but the cellulose sugars are linked in a fashion that is not easily broken down by naturally occurring organisms and enzymes. In contrast, starch is composed mainly of six-carbon sugars that are easily broken down by native organisms. Thus, cellulose feedstocks, while used widely for paper manufacture and as a boiler fuel, have not been used for the production of ethanol, and leaves, stalks, and other plant parts are considered "agricultural wastes."

If a practical path to produce ethanol from cellulose could be found, both the premium energy balance argument and the economic reasoning discussed above for corn-derived ethanol would be turned on their heads. For many years, therefore, finding a way to break down cellulose so that it can be converted to ethanol or other useful products has been an important technical challenge. In the past few years, BC International Corporation (BCI) has put forward a practical scheme for accomplishing this goal. BCI utilizes hot sulfuric acid solutions to separate the cellulose five- and six-carbon sugar chains from the lignin. The key technical advance on which the BCI process is based is the production of a genetically engineered organism that is able to digest the C-5 sugar chains and produce ethanol. (The new organism was developed by Professor L.O. Ingram, a microbiologist at the University of Florida).

The BCI process uses agricultural waste as feedstock. It requires more process heat for the fermentation because of the energy needed to separate the cellulose and because of the lower activity of the bio-engineered organism. However, because agricultural residues are the feedstock for the process, a portion of the process heat (we will assume about half of it) can come from this source. And because the residues are a waste product, the premium energy required to grow the primary crop need not be included. A simplified process flow sheet is shown in Figure 2.2. It is immediately apparent that the process has a favorable premium energy balance:

$$\text{Premium energy gain} = 84{,}000 - 40{,}000 = +44{,}000 \text{ BTU per gal.}$$

The economics of the cellulose-based ethanol production technology are also promising, principally because the expense of buying corn is avoided. If we assume the price of corn is $2.50 per bushel and 0.4 bushels of corn are required to produce a gallon of ethanol, about $1 per gal of cost is avoided. Because the total cost of producing ethanol from corn is about $1.50 per gal, this cost advantage is considerable. However, there are several

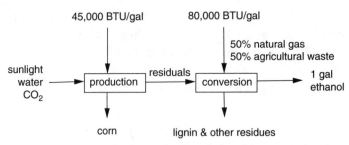

Figure 2.2. Flow diagram for the production of ethanol from agricultural waste.

costs that partially offset this advantage: (1) the cost of collecting and transporting the agricultural residue to the plant, (2) the cost of the bio-engineered microorganism that will be greater than the cost of the native enzymes and yeast used in corn fermentation, and (3) the higher plant capital cost attributable to the fact that the cellulose-based process has more steps, more severe operating conditions, and takes place more slowly. Nevertheless, the economics of the cellulose-based process could be quite favorable compared to conventional corn-based production of ethanol.

Not surprisingly, there has been considerable interest in this new technology. For example, a recent article in *Foreign Affairs* pointed out the promise of this new approach and the advantages it has in terms of energy efficiency, environmental impact, and reduced oil imports.[9] The important point here is that a simple technical and economic analysis demonstrates why one should be skeptical of corn-based ethanol production for gasohol, but supportive of cellulose-based ethanol production.

The gasohol case illustrates three important points:

(1) a careful analysis of the outputs and inputs to a process will show whether a program achieves the objective for which it is designed. In this case, gasohol does not achieve the objective of displacing premium petroleum fuel – or does so only barely.
(2) Tax credits adopted for a flawed objective can distort the underlying economics so that the program seems to be economically viable.
(3) An astute analysis can point to technical innovations that reverse or compensate for the undesirable characteristics of a process. Here we see that while ethanol from corn may be a bad idea, ethanol from cellulose and agricultural waste is certainly not.

This conclusion points the way to government support that may make sense – support for cellulose-based ethanol production rather than support for corn based gasohol that does not make sense.

---

[9] R. James Woolsey and Richard Lugar, "The New Petroleum," *Foreign Affairs*, 78(1), p. 88–102 (1999).

3

Solar Thermal, Windpower, and Photovoltaic Technologies

Many experts believe that nonrenewable fuels, in particular oil and gas, will eventually become so scarce and therefore so expensive that they will no longer be practical large-scale energy sources. Moreover, the use of coal and other fossil fuels imposes major environmental burdens. Therefore, it is prudent to develop energy technologies based on renewable energy sources and introduce them commercially if and when they become economically competitive.

There are many renewable energy technologies to consider. First are those that rely on natural terrestrial forces: wind, geothermal, hydropower, and tidal power. Second, there are technologies that rely directly on solar energy. These include solar hot water heating, solar thermal electric conversion (either in solar "power towers" or in the more exotic form of solar power satellites), and photovoltaics. Some terrestrial energy sources can be regarded as indirect forms of solar energy. For example, solar ponds and ocean thermal energy conversion (OTEC) rely on solar-heating-induced temperature gradients. Similarly, solar heating of the atmosphere drives the winds. Biomass, another important renewable energy source, can also be regarded as an indirect form of solar energy.

We shall not analyze all these technologies in this chapter. Rather our purpose is to describe a process for evaluating and comparing competing technologies. We consider three important renewable energy technologies in some detail: solar hot water heating, wind energy, and photovoltaics. In each case the task is to evaluate the technical and economic feasibility of substituting the renewable technology for traditional energy sources.

There is considerable public and political interest in renewable energy technologies at present. Our goal in this chapter is to encourage an objective, unemotional view of what can be expected from these technologies. Of course, almost everyone would welcome the invention of a new source of energy that did not involve the significant environmental burdens and economic costs of current petroleum, coal, and nuclear-based technologies. Our task here is to analyze renewable technology options and their economic costs relative to these traditional energy sources. The public enthusiasm for renewable energy should not be ignored, but neither should it push us into costly courses of action that raise expectations but do not deliver on their promise. We will see in Chapter 7, on nuclear power, that it is not easy to manage technology development when there is public mistrust of the technology. Here we shall see that it can also be difficult to manage the development of a technology for which there is torrid public enthusiasm. In both cases, dogma, fashion, or good intentions are no substitute for careful analysis.

Figure 3.1. Rooftop solar collectors on a house in Golden, Colorado.

SOLAR HOT WATER HEATING

Hot water heating is one of the simplest applications of solar energy. The sun's energy is used to heat water for residential or commercial use. The technology is intended to substitute for gas or electricity-powered hot water heaters. In many areas of the world, especially in tropical climates, solar hot water heating is used extensively. Figure 3.1 shows a house in Golden, Colorado equipped with solar collectors.

For certain applications, the solar heating system focuses solar radiation onto a vessel or pipe containing a heat transfer fluid. Solar power towers using liquid sodium as the working fluid can generate temperatures high enough for industrial process heat applications. Figure 3.2 shows a parabolic trough solar collector used for hot water heating at a prison in Colorado.

The Sun's Energy

The sun behaves as a black body radiator at a temperature of about 6,000 degrees Kelvin. The Planck distribution law describes the frequency distribution of radiation emitted by a black body at absolute temperature T:

$$E(\nu) = \frac{8\pi h\nu^3}{c^3} \frac{\exp(-h\nu)/kT}{(1 - \exp(-h\nu)/kT)},$$

Figure 3.2. Parabolic trough solar collector for hot water heating at a prison in Adams County, Colorado.

where ν is the frequency, h is Planck's constant, c is the speed of light, T is the temperature, and k is Boltzmann's constant. The maximum of the energy distribution occurs at a frequency $\nu_{\max}$, given by:

$$\frac{h\nu_{\max}}{kT} = \frac{hc}{kT\lambda_{\max}} = 2,$$

where $\lambda_{\max}$ is the wavelength at the maximum of the energy distribution. The black body emits radiation at a rate proportional to T^4:

$$q \equiv \text{flux} \propto \overline{E} = \int_0^\infty d\nu\, E(\nu) \propto T^4.$$

In the Earth's mid-latitudes, the radiation reaching the earth's surface on a cloudless day at noon is about 1 kw per m². The solar radiation ranges in wavelength from 0.2 microns to 4.0 microns (1 micron $= 10^{-6}$ meters). At wavelengths longer than about 0.67 microns ($h\nu < 1.8$ eV), the radiation falls in the infrared region; at wavelengths shorter than 0.4 microns ($h\nu > 3$ eV), the radiation is in the ultraviolet. The energy intensity of solar radiation peaks at approximately 0.8 microns (equivalent to $\sim$1.5 eV).

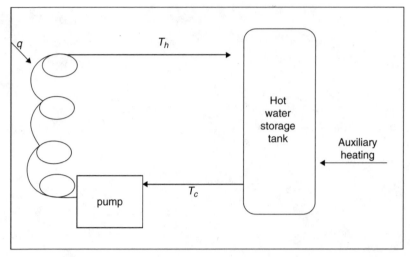

Figure 3.3. Schematic of residential solar hot water heating system.

A Practical Design and Calculation

A typical U.S. family of four in a three-bedroom house uses about 100 gal of hot water (at 150°F) per day. A solar thermal system designed to provide this hot water is shown schematically in Figure 3.3. Cold water at temperature T_c is pumped to solar collectors where it is heated to temperature T_h. This hot water is stored in a tank for later use. Auxiliary heating is provided for when there is no sun.

We assume that the cold water enters the solar collector at 40°F and that the hot water in the storage tank should be at 150°F. The daily BTU requirement for hot water heating is:

$$\text{BTUs required/day} = (100 \text{ gal per day}) \times (8.33 \text{ lbs. per gal})$$
$$\times (1 \text{ BTU per lb.°F}) \times (150 - 40)(°F)$$
$$= 91{,}630 \text{ BTU per day.}$$

Thus the required heat load is roughly 90,000 BTU per day. Assuming a 50% collector efficiency this requires 180,000 BTU per day of solar radiation.

How much solar heat does the sun deliver each day? This obviously depends on both season and location. Typical values are given in the following table:

Solar BTUs delivered per square foot per day

	January	June	Average
Boston, MA	500	2,000	1,000
Tucson, AZ	1,000	3,500	2,000

If we decide to meet 50% of the required annual heat load from solar energy (for reasons that will be explained shortly), we will require about 90 sq ft of hot water solar collector area per home in Boston and about 45 sq ft in Tucson. Averaged over the year, these systems will deliver about 50 gal of hot water per day at a temperature of 150°F, assuming a feed water temperature of 40°F. The systems thus provide, on average:

$$Q = V.\rho.\Delta T.c \simeq 45,000 \text{ BTU of heat per day,}$$

where V, the volumetric flowrate is 50 gal per day, the density of water $\rho = 8.33$ lbs. per gal, the temperature increase $\Delta T = (150 - 40)°F$, and the specific heat of water $c = 1$ BTU/°F-lb.

Economic Analysis

Is it worthwhile to switch from gas to solar heat for the hot water this home requires? In order to answer this question, we must compare the cost of the two alternatives.

For the solar hot water heating system, we assume that once the system is installed (and paid for), the continuing operating costs are negligible, that is, the cost of electricity for the pump and the cost of maintenance are assumed to be very small. We also ignore the cost of the hot water tank, because this must be purchased for either the solar or conventional gas hot water heating system. The cost estimates for the main components of the solar system are presented in the following table.

System cost	Boston	Tucson
1. Panels (@$17/ft^2)	$1,530	$765
2. Piping	500	500
3. Pump & controls	100	100
4. Installation	500	500
Total	$2,630	$1,865

Installation of the solar hot water heating system delivering an average of 45,000 BTU per day (or, equivalently, 16.4 million BTU per year) will thus cost about $2,600 in Boston and $1,900 in Tucson. How does this compare to the cost of conventional hot water heating by gas? To make this comparison, we must weigh a one-time investment in the solar heating system against the recurring expense of purchasing the gas. (We assume that there is no capital cost for the gas system, because the gas burner is integrated into the hot water storage tank.) One way to make this comparison is to apportion the capital cost of the solar heating system uniformly to each year of its operating life, and then compare this "annualized" cost with the annual cost of the gas. But simply dividing the capital cost by the number of years of life of the solar hot water system – ten years, say – would understate the true cost of the investment, because this would ignore the interest cost of the invested capital. To see why, imagine that the homeowner borrows all the money from the bank to buy the solar heating system. Let us further assume that the loan is for a fixed ten-year term, at a constant interest rate of 10% per year. Each year, the homeowner must pay off part of the loan principal and pay interest on the portion of the principal that is still outstanding. To make the comparison easier, let us further assume that the terms of the loan require the borrower to make equal annual payments

Table 3.1. Threshold delivered price of natural gas above which residential solar hot water heating is economical ($/MCF)

Loan interest rate $r \times 100$ (%/yr)	Annual capital charge rate, ϕ (%/yr)	Threshold price of gas, p^* ($/MCF)	
		Boston ($I = \$2,630$)	Tucson ($I = \$1,865$)
3	11.7	15	10.5
6	13.6	17.4	12.5
10	16.3	20.9	14.8

to the bank throughout the life of the loan – that is, the annual "debt service," the sum of the principal repayment and the interest owed on the remaining principal, is the same in each year. For this case it is straightforward to show (the derivation is given in the Appendix to this chapter) that for a loan of I dollars made at an interest rate of r per year (or, equivalently, $100r$ percent per year) and for a term of N years, the annual loan payment is:

$$\text{Constant annual loan payment} = I \left[\frac{r(1+r)^N}{(1+r)^N - 1} \right]. \tag{1}$$

The term in square parentheses in equation (1) is called the annual capital charge rate, ϕ. As already noted, the annual loan payment is comprised partly of interest on the outstanding principal and partly of principal repayment. In the early years of the loan, interest accounts for the lion's share of the payment; toward the end of the term, most of the payment goes toward repaying the principal. (The trajectory of interest and principal repayments for a ten-year loan made at 10% per year is given in the Appendix.)

We can now calculate the maximum allowable price of natural gas, p^*, in dollars per thousand cubic feet ($ per MCF), above which the annual cost of purchasing gas exceeds the annual loan payment on the solar heating system:

$$I\phi = p^* Q,$$

where Q is the annual gas requirement (in MCF).

The heat content of natural gas is about 1 million BTU per MCF. If we assume an 80% heating efficiency for the gas, the annual gas requirement is given by:

$$Q = (16.4 \times 10^6)/0.8 \text{ (BTU per yr)} \times 10^{-6} \text{ (MCF per BTU)}$$
$$= 20.5 \text{ MCF per yr.}$$

The crossover price of natural gas above which solar heating is economic in Boston and Tucson is shown in Table 3.1 for different values of the interest rate, where we have again assumed that the solar system is financed with a ten-year loan.

As expected, solar hot water heating is competitive at a lower gas price in sunny Tucson than in Boston. But residential gas prices are considerably higher in Boston

Table 3.2. Threshold delivered price of natural gas ($/MCF) above which residential solar hot water heating is economical, assuming a 50% tax credit

Loan interest rate r (%/yr)	Annual capital charge rate, ϕ (%/yr)	Threshold price of gas, p^* ($/MCF)	
		Boston ($I = \$1,315$)	Tucson ($I = \$940$)
3	11.7	7.5	5.25
6	13.6	8.7	6.25
10	16.3	10.45	7.4

than in Tucson because of the higher cost of transporting gas to the northeastern United States. If the difference in delivered gas prices between the two cities is large enough, solar hot water heating could in principle be economical in Boston and not in Tucson.

In fact, the average price of natural gas delivered to residential consumers in Massachusetts during 2001 was $13.35 per MCF, and $10.34 per MCF in Arizona. So, for a typical interest rate (of 6% per year or more), we conclude on the basis of this analysis that solar hot water heating is not economical in either location.

However, this analysis has not taken into account the effect of tax credits for residential solar installations, which in some instances have been granted for up to 50% of the installation cost. Such credits reduce the effective investment cost by 50%, which would be enough to make solar hot water heating economical for homeowners in some parts of the country, as shown in Table 3.2. Of course, the cost to society of the solar option would not have changed – it is the taxpayer who is effectively paying the difference. There may be good reasons for such a subsidy, for example: (1) if the environmental costs of using the competing fuel (in this case natural gas) are not internalized; (2) if the market is overestimating the future availability of natural gas; or (3) if the subsidy will stimulate technological change in the design and manufacture of solar collectors that will result in lower costs in the future. We will discuss such issues later in this chapter, and in subsequent chapters, too. The main points to emphasize at this stage are that subsidies can significantly affect the competitiveness of individual energy sources, renewable as well as conventional, and that the standard for applying subsidies should be consistent across different technologies.

Why is the System Designed for Only a Fraction of the Load?

At any given location, the solar flux varies with season. If the system were designed to satisfy 100% of the required heat load throughout the year, the solar collector area would have to be large enough to meet the load during the part of the year when the solar flux is smallest. For the rest of the year, the system would produce excess hot water. Reducing the collector area would lessen the cost, but would also mean that the heat load would only partly be met when the flux was lowest, requiring the purchase of backup heating during that time of the year. Reducing the collector area further would lengthen the

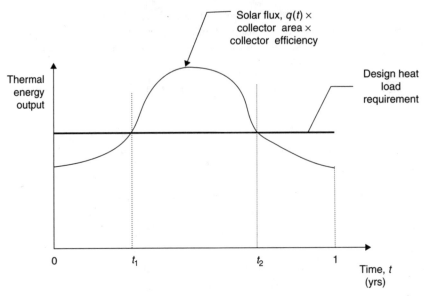

Figure 3.4. Energy output of solar hot water system relative to design heat requirement as a function of time during the year.

interval during which there would be insufficient hot water to meet the load, further increasing the requirement for backup energy. Beyond a certain point, the collector area would be so small that there would be no time of the year when it would be capable of meeting 100% of the load. In general, there is an optimal size for the collector, which is determined by the economic tradeoff between the capital cost of the collector and the cost of purchasing auxiliary heat energy. The situation is shown in Figure 3.4. For times $t_1 < t < t_2$, 100% of the heat load is met; for times $t < t_1$ and $t > t_2$, backup energy is required. In the example discussed previously, the optimal collector was assumed for simplicity to deliver 50% of the required load.

WIND ENERGY

For centuries wind was used throughout the world as an important source of mechanical power for pumping water, grinding grain, and other applications. In later times wind power fell into disuse in most advanced economies. Its intermittent nature made it uncompetitive with the round-the-clock availability of steam power from inexpensive fossil fuels. Recently, however, the use of wind power, especially for electricity generation, has been increasing in both the advanced and developing countries. Its contribution to the world's supply of electricity is still small (less than 1%), though in some parts of the world it is growing rapidly. The total wind energy resource is very large. About 1% of the incoming solar flux goes to drive the winds, or about 1,200 terawatts (1 terawatt = 10^{12} watts). This is roughly 100 times the current global rate of energy use. But only a very tiny fraction of this energy could ever be captured economically.

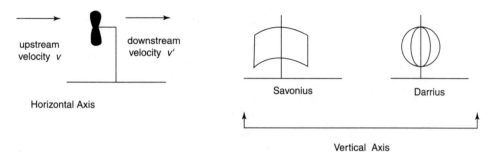

Figure 3.5. Examples of different wind turbine types.

Technical Performance of Wind Machines

Over the years many different types of wind turbines have been proposed with both horizontal and vertical axes of rotation. Some of these are illustrated in Figure 3.5. A Darrius-type wind turbine at the large wind farm at Altamont Pass, California is shown in Figure 3.6.

The basic concept of wind power is straightforward. The wind's kinetic energy is converted into the rotational energy of the turbine blades, which is then converted to electrical energy by a generator. The amount of power produced by a wind machine depends both on the strength of the wind and the size of the windmill blades.

The power output of the windmill is estimated by considering the kinetic energy per unit time delivered to the blades. The volume of air ΔV delivered in time period Δt is given by:

$$\Delta V = v A \, \Delta t,$$

where A is the area of the disc described by the rotor blades normal to the wind direction and v the incoming wind speed.

The kinetic energy in a parcel of air of unit volume travelling at speed v

$$= (\rho v^2/2),$$

where ρ is the density of the air. Thus, the kinetic energy delivered in time period Δt

$$= (\rho v^3/2) A \, \Delta t.$$

Therefore, the power, P, delivered is

$$P \propto \rho v^3 A/2.$$

The important points here are that the power P is proportional to v^3 and is also proportional to the area A swept by the blade, that is, $P \propto D^2$ where D is the blade diameter.

Not all of the power in the wind can be used by the wind machine, since this would require the wind velocity downstream of the turbine to fall to zero. When the constraints of pressure and mass continuity are taken into account, the maximum theoretically

Figure 3.6. A Darrius-type vertical axis wind turbine.

achievable power is

$$P = c_d \rho v^3 D^2,$$

where the constant of proportionality, $c_d = (16\pi/27 \times 8)$.

For a typical value of air density, the theoretical maximum power in kilowatts is:

$$P(\text{in kw}) = (0.022) \times (v \text{ in mph})^3 \times (D \text{ in hundreds of feet})^2.$$

For example, if:

$$v = 30 \text{ mph and } D = 100 \text{ ft, then } P = (0.022)(30)^3(1)^2 = 590 \text{ kw}.$$
$$v = 20 \text{ mph and } D = 300 \text{ ft, then } P = (0.022)(20)^3(3)^2 = 1584 \text{ kw}.$$

This is the power delivered by the wind to the turbine. The turbine turns a generator and produces electricity with an overall conversion efficiency of about (2/3).

So,

$$P_{out}(\text{kw}) = (0.014) \times (v \text{ in mph})^3 \times (D \text{ in hundreds of ft})^2.$$

Constructing large wind turbines is a technical challenge. One reason is that the speed at the blade tip increases with diameter and this places terrific stress on the blades. For example, a 66-meter diameter blade (roughly equivalent to the wingspan of a jumbo jet) rotating with a frequency, f, of 1 revolution per second has a blade tip speed of

$$v \text{ (tip)} = \omega \frac{D}{2} = 2\pi f \frac{D}{2} = 207 \text{ m/sec}.$$

The system can also fail if the support tower and/or the rotor mount are not rigid enough. This can result in tilting of the mount and rotor axis, inducing rotational instabilities which can cause the blades to shake back and forth and eventually break off. Modern wind machines are typically designed such that when the wind velocity exceeds a certain level (the cut-out velocity), power is no longer extracted in order to protect the system from excessive stresses.

How Big a Wind Machine Should One Build? We can gain insight into this question by asking how the cost of the wind energy system scales with size. This type of scaling analysis is often useful. First, we note that wind speed varies with height above the ground surface, Z, according to a power law, $v(Z) \propto Z^{1/7}$

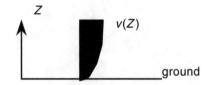

because D is roughly proportional to Z, we have that $P \propto Z^2 v^3$, that is, $P \propto Z^{17/7}$.

The mass, M, of the tower is proportional to the volume of the tower, which scales roughly as Z^3. Thus, the mass of material per unit power produced varies according to:

$$(M/P) \propto Z^{4/7}.$$

This says that the mass of material needed to build a wind turbine increases more than proportionally with the size of the rotor. Thus, if material cost is the major factor in the cost of wind energy, the turbine size should not be too large. (Of course, there are other contributors to the total cost of the system, including the electrical generator, whose scaling behavior can be expected to be different.)

Modern windmills typically range in size from 25 kw to 2 mw. A 225 kw horizontal axis wind turbine installed on San Clemente Island, California in 1998 is shown in Figure 3.7. The rotor diameter is approximately 30 meters, and the area swept by the rotors is approximately 700 m^2. The wind machine starts producing power at a wind speed of approximately 9 mph, and continues to do so up to a wind speed of 56 mph. The 250 kw

Source: Courtesy National Renewable Energy Laboratory, Photographic Information Exchange.

Figure 3.7. Wind turbine at San Clemente Island, California (225 kw).

two-bladed wind turbine shown in Figure 3.8, also located in California (in the San Gorgonio Pass near Palm Springs), uses aircraft-style aileron controls to smooth out energy spikes in high wind conditions.

The Quality of the Wind Resource is Very Important. A good wind farm site is one where the wind blows strongly and steadily. Because the available wind power increases as the

Source: Courtesy National Renewable Energy Laboratory, Photographic Information Exchange.

Figure 3.8. Two-bladed turbine at San Gorgonio Pass, California (250 kw).

cube of the wind speed, even small fluctuations in wind speed cause large variations in available power. A "wind duration curve" at Kahuku on the island of Oahu in Hawaii is shown in Figure 3.9. At this site the trade winds are very steady and above average in strength, and the site is one of the world's best for a wind farm.

Examples of other installations of wind machines are given in the table below:

Machine	$<v>$ Rated	Diameter	$P(<v>)$	P(actual)
	(mph)	(feet)	(kw)	(kw)
U.S. Wind	26	30	22	25
Grandpa's Knob, VT	30	175	1,158	1,250
Boeing Block Is., RI	28	300	2,756	2,800

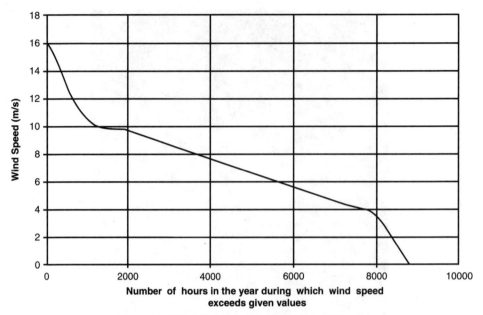

Figure 3.9. Wind duration curve for Kahuku, Hawaii.

Note that according to the table the actual power is higher than the theoretical maximum power predicted based on average wind speed.[1]

Several authors have presented valuable surveys of wind energy technology.[2]

Economics of Wind Energy

A block diagram of a wind system is shown in Figure 3.10. The system consists of the rotor, the electrical generator and gearing, the support tower, and the power conditioning

---

[1] How can it happen that the actual power is greater than the theoretical maximum? The explanation has to do with the variability of the wind speed. If the windmill can always turn into the wind direction, then: $<P(v)> > P(<v>)$. If $f(v)$ is the distribution of wind speeds at a site and δv is the deviation of the instantaneous wind speed from the average $\delta v = v - <v>$, then:

$$\frac{<P(v)>}{c_d D^2} = <v^3> = \int_0^\infty dv f(v) v^3 = \int_0^\infty dv f(v) <v>^3 \left[1 + \frac{\delta v}{<v>}\right]^3.$$

In the limit of small deviations from the average, one finds:

$$\frac{<P(v)>}{c_d D^2} - \frac{P(<v>)}{c_d D^2} = <v><(\delta v)^2> > 0.$$

This is an illustration that the function of the average is not equal to the average of the function. Sometimes this difference works in one's favor and sometimes not.

[2] See, for example, Frank R. Eldridge, *Wind Machines*, Van Nostrand Reinhold, New York, 1980 (2nd edition); J. G. McGowan and S. R. Connors, "Windpower – A turn of the century review," *Annual Reviews of Energy and the Environment*, 25, Annual Reviews, Palo Alto, CA, 2000; Bengt Sorenson, "History of, and recent progress in, wind energy utilization," *Annual Reviews of Energy and the Environment*, 20, Annual Reviews, Palo Alto, CA, 1995; G. Thomas Bellarmine and Joe Urquhart, "Wind Energy for the 1990s and Beyond," *Energy Conversion Management*, 37 (12), pp. 1741–1752 (1996).

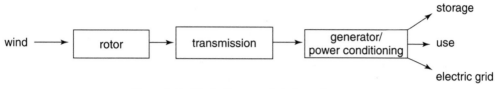

Figure 3.10. Block diagram of wind energy system.

equipment. The generator works best when it runs at constant rotational speed, producing uniform output voltage and current. The controls and transmission must convert variable wind input to smooth power input to the generator. Wind turbines are dynamically fragile, and catastrophic failure (especially of rotor blades) is not unknown. Cost estimates are uncertain because the reservoir of experience on which to base operating and maintenance cost projections is still relatively small for the latest generation of machines, especially in the later years of life. A low capacity factor (due to wind variability) is a drawback for electric power applications (although not for the agricultural application of pumping water), and either a backup source of electricity or an energy storage device is a vital part of a wind energy system.

How Do We Know if Wind is an Economically Attractive Power Source? We will adopt the perspective of a private investor who is considering putting his or her money into a new business venture. How would such a person evaluate the economics of wind (or indeed any other sort of business opportunity)?

For specificity, assume the wind project consists of a wind farm consisting of 100 machines, each with a rated capacity of 25 kw electric, at a site with favorable wind characteristics. The local utility agrees to buy the electricity produced by the wind farm (either because it needs the energy or because it is told to do so by the utility's regulatory commission as part of a scheme to promote renewable energy resources). This means that the project need not be concerned either with energy storage or with distribution to end-users; the wind farm is simply connected to the utility's grid. (Of course, the utility may need to arrange for standby or backup electricity capacity in order to compensate for the inherent variability of the wind energy). We will also assume, initially, that the entrepreneur finances this project entirely with her own funds.

In considering this investment, we make certain assumptions about the project's performance and economics. We assume that the project requires site preparation and engineering work costing $300,000 and that the wind machines cost $25,000 each. Further, the annual operating and maintenance (O&M) cost per wind machine is $1,300. Also, even though the wind regime at the site is favorable, the capacity factor of the machines (defined as the ratio of actual electrical energy produced during a typical year to the energy generated if the turbine operated continuously at its rated power) is 35%. Finally, we assume that the wind machines will operate for 13 years, because that is the length of time for which the utility has agreed to pay a price of $.075 for each kwhr of power delivered – a fairly generous offer. Windpower also currently enjoys a 1.5¢ per kwhr federal renewable energy production tax credit, which provides an additional

Cost & Revenue Streams (undiscounted)

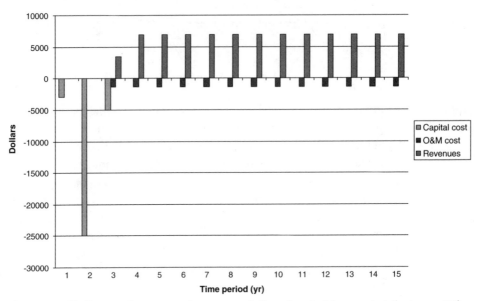

Figure 3.11. Undiscounted expense and revenue cash flows for wind farm project (basis: one 25 kw wind machine).

stream of revenue to the entrepreneur. With these assumptions, the annual revenues per wind turbine

$$= 25 \text{ (kw)} \times 365 \text{ (days/yr)} \times 24 \text{ (hrs/day)} \times (0.075 + 0.015)(\$/\text{kwhr}) \times 0.35$$
$$= \$6,899/\text{yr}.$$

We assume that it takes a year to prepare the project and a further year to procure and install the wind machines. In year three, the first year of operation, final site preparation and interconnection costs of $5,000 per windmill are incurred. Also in this first year, there is only a half-year of operating revenue. The resulting cash flows per wind machine are indicated in Figure 3.11. (The convention in these cash flow diagrams is to put cash outlays (expenses) below the time line and cash inflows (receipts) above.)

At first glance, the way to determine whether the project is profitable would simply be to sum all the expenditures and sum all the income streams, subtract the former from the latter, and see whether the answer was positive or negative. But we cannot do this, because the value of each individual cash flow depends on the point in time at which it occurs.

The basic principle here is quite simple: A dollar received today is worth more than a dollar received a year from today because today's dollar can be productively invested and earn a return. For example, $1 invested in a savings account today would be worth $(1 + r)^n$ n years into the future, assuming a compound interest rate of r per year. Conversely, a dollar received n years in the future is worth less than a dollar received today because today's dollar can be invested and earn interest until then. The amount received today that would be equal in value to $1 received n years from now is $\$(1 + r)^{-n}$. The two sums of money, $\$(1 + r)^{-n}$ today and $1 n years from now, are economically equivalent: One would be indifferent about choosing between them. The corollary is clear: You cannot add or subtract cash flows occurring at different points in time without first converting them to a common time base.

Converting a cash flow occurring today to its equivalent value at some future date is called finding that cash flow's future worth. The future worth in n years, F, of a present cash flow, P, for an interest rate r, is given by:

$$F = P(1 + r)^n.$$

The multiplier in the above equation, $(1 + r)^n$, is called the future worth factor or compound amount factor. Conversely, the equivalent value today, P, of a future cash flow F is called the present worth of that cash flow. The conversion is

$$P = F(1 + r)^{-n},$$

where the factor $(1 + r)^{-n}$ is called the present worth factor. Calculating the present worth of a future cash flow is called discounting the cash flow to the present, and the interest rate used in the calculation is referred to as the discount rate.

The correct way to determine the profitability of the wind farm project, therefore, is first to convert all of the cash flows in Figure 3.11 to a common time base. In this case we discount the cash flows to the beginning of the project, using a discount rate of 7% per year (we could equally well have calculated the future worth of all the cash flows at the end of the project operating life.) Figure 3.12 shows the net cash flow in each year, along with its discounted value or present worth at time 0. Note the decline in the discounted values of the cash flows in the later years of the project. The present worths of these net cash flows are also shown in Table 3.3. The sum of all these present worths, the net present value (NPV) of the wind machine, is $9,331. The overall 100 wind turbine project therefore has a present value to the investor of $933,100.

The Role of Risk

In principle, any project with a positive net present value is worth undertaking. Certainly no one would undertake a project with negative NPV. Entrepreneurs can in general be expected to undertake projects with high NPVs. But these values are not certain to be realized. The assumptions we made in this calculation were reasonable, but they were only assumptions. Note also in Table 3.3 that the cumulative NPV of the investment in the wind turbine does not turn positive until year 11, which is plenty of time for things to go wrong. In other words, there is risk – the future may not turn out as projected. Depending on the magnitude of the risk, the entrepreneur may elect not to pursue the project.

Using Probabilities to Assess Risk. As an illustration – a particularly severe illustration – of risk, suppose that there is a probability, p_1, that the utility will renege on its contract

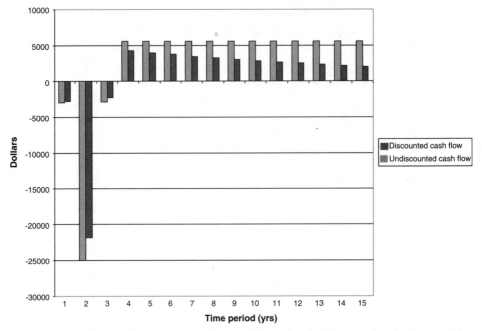

Figure 3.12. Undiscounted and discounted net cash flows for wind farm project (basis: one 25 kw wind machine; $r = 7\%/\text{yr}$).

after the wind farm has been built, but before it goes into service. If this were to happen, the net present value of a wind turbine would be about $(-\$29,000)$, that is, the wind machine would lose this amount of money. The expected value of the project should then be evaluated as:

$$\text{Expected value} = p_1(-\$29,000) + (1 - p_1)(\$9,330).$$

The result depends on the probability of default. Note that if this is greater than $p_1 = 9,330/38,330 = 0.24$, the expected value of the project is negative.

Sensitivity Analysis as a Means of Assessing Risk. The likelihood of a complete default by the utility is quite low. On the other hand, there are many other possibilities that would negatively affect the economics of the project (as well as some that would have a positive effect.) "Pro forma" project estimates (that is, those made in advance of the project) almost never turn out to be completely accurate. Therefore, a disciplined method for examining the consequences of differing assumptions about the future is of great importance in evaluating investment decisions. The most straightforward approach is to calculate the NPV of the project for a series of parameter values above and below the value considered most likely. The result gives the sensitivity of project outcome to alternative assumptions about key project parameters.

As an example, consider the variation in the present value of the wind project we are considering if the interest rate changes by 0.5% above and below the base case of 7%. The results of these calculations are given in the table below. Note the sensitivity

Table 3.3. Present value of wind machine (cash flows in dollars; interest rate $= 7\%/\mathrm{yr}$)

Year n	Expenditures	Revenues	Present worth of net cash flow in year n	Cumulative NPV after n years
1	−3,000		−2,804	−2,804
2	−25,000		−21,836	−24,640
3	−6,300	3,450	−2,327	−26,967
4	−1,300	6,900	4,271	−22,696
5	−1,300	6,900	3,992	−18,704
6	−1,300	6,900	3,731	−14,974
7	−1,300	6,900	3,486	−11,487
8	−1,300	6,900	3,258	−8,229
9	−1,300	6,900	3,045	−5,183
10	−1,300	6,900	2,846	−2,337
11	−1,300	6,900	2,660	322
12	−1,300	6,900	2,486	2,808
13	−1,300	6,900	2,323	5,131
14	−1,300	6,900	2,171	7,303
15	−1,300	6,900	2,029	9,332

of the results: A 10% increase in the interest rate, from 7% to 7.7%, reduces the NPV of the project by 18% from $9,332 to $7,680 per wind machine.

Interest rate (%)	Net present value ($)
5.5	13,358
6.0	11,936
6.5	10,336
7.0	9,332
7.5	8,141
8.0	7,018
8.5	5,959

Many other cases should be considered in order to gain an understanding of which aspects of the project are most critical to a favorable economic outcome. As a second example consider the variation of NPV with the sale price of electricity:

Electricity sale price(¢/kwhr)	Net present value ($)
6	1,407
6.5	4,049
7	6,690
7.5	9,332
8	11,973
8.5	14,614

The result shows that the project economics are even more sensitive to assumptions about the electricity price. A 1 cent reduction below the base case of 7.5¢ would reduce the NPV by 56%.

The Concepts of Internal Rate of Return and Debt Financing

The investor wants to know how profitable the project is. The NPV is one such measure. Another that is popular with investors is the internal rate of return (IRR). The IRR, denoted here by the symbol R to distinguish it from the market interest rate, r, is the discount rate at which the net present value of the project is equal to zero. The IRR is determined from the implicit equation:

$$0 = \sum_{n=1}^{N} \frac{C_n}{(1+R)^n}, \qquad\qquad (2)$$

where C_n is the net cash flow in time period n.

One way to interpret the IRR is that it is the highest value of the market rate of interest, r, at which the project remains economically viable. If $r < R$ the project would be economic in the sense that one could borrow money at r, invest it in the project, repay the loan, and make a profit. If, on the other hand, $r > R$, the project would not be economic in this sense.

For the wind project under consideration the IRR is found from equation (2) above to be $R = 0.12$, that is, 12%. (The solution is easily found using standard spreadsheet software such as Microsoft Excel.) Because the (assumed) market interest rate of 7% is less than this, it would be worthwhile to borrow money in order to invest in this project. As long as $r < R$, this is an interesting strategy.

For example, consider borrowing 80% of the cost of purchasing the wind machines (that is, $20,000 per machine). For simplicity, we assume that the bank is willing to lend this money at a rate of 7% per year. The bank will be willing to make such a loan if it thinks the project has a good chance of succeeding and because its money will be secured by a tangible asset, the wind machines. If the project starts to go bad, and the actual cash flows are less than projected, the entrepreneur may be forced to default on his loan payments. At that point the terms of the loan give the bank the option to step in, foreclose on its loan, seize the wind farm, and either operate the project itself or, more likely, sell it in order to recover its capital. Depending on the sale price the bank might or might not recover all of its capital in this scenario. Because it carries a residual risk, the bank will in practice charge an interest rate that is higher by some increment – a "risk premium" – than the 7% market rate. If the bank has only lent a small fraction of the total investment cost, it stands a better chance of recovering all its capital if the project goes bad. Its risk of not doing so increases as the fraction of the total investment cost accounted for by the loan rises. For this reason, banks are only rarely willing to lend 100% of the investment cost. (Another reason is that the bank likes to know that the entrepreneur's own money is invested in the project and so also at risk, like the bank's funds.)

Table 3.4. Wind project cash flows, with and without 7%/yr debt financing of 80% of the wind turbine cost (basis: single wind machine; cash flows in dollars; interest rate = 7%/yr)

Year	Without debt financing			With 80% debt financing (@7%/yr)			
	Net cash flow (1)	PV of net cash flow (2)	Cumulative NPV (3)	Loan payment (4)	Net cash flow (5) = (1) + (4)	PV of net cash flow	Cumulative NPV
1	−3,000	−2,804	−2,804		−3,000	−2,804	−2,804
2	−25,000	−21,836	−24,640	20,000	−5,000	−4,367	−7,171
3	−2,851	−2,327	−26,967	−2,393	−5,244	−4,280	−11,451
4	5,599	4,271	−22,696	−2,393	3,205	2,445	−9,006
5	5,599	3,992	−18,704	−2,393	3,205	2,285	−6,720
6	5,599	3,731	−14,974	−2,393	3,205	2,136	−4,585
7	5,599	3,486	−11,487	−2,393	3,205	1,996	−2,588
8	5,599	3,258	−8,229	−2,393	3,205	1,866	−723
9	5,599	3,045	−5,183	−2,393	3,205	1,744	1,021
10	5,599	2,846	−2,337	−2,393	3,205	1,630	2,650
11	5,599	2,660	322	−2,393	3,205	1,523	4,173
12	5,599	2,486	2,808	−2,393	3,205	1,423	5,597
13	5,599	2,323	5,131	−2,393	3,205	1,330	6,927
14	5,599	2,171	7,303	−2,393	3,205	1,243	8,170
15	5,599	2,029	9,332	−2,393	3,205	1,162	9,332
	IRR = 12%/yr				IRR = 18%		

How does the loan affect the project economics from the entrepreneur's point of view? Table 3.4 shows the project cash flows with and without the loan. (As with the example of the solar hot water heater, the loan is assumed to be paid off in equal installments, in this case over a thirteen-year term. The annual loan payment is given by equation (1) above.)

Table 3.4 contains some interesting insights about the possibilities of investment transactions. Without the loan, the investor puts in $33,000 of his own funds per wind machine and realizes a positive net present value of $9,332, with eleven years required before the NPV turns positive. The IRR of this investment is 12%. With the loan, the NPV of the project remains the same, but in this case the investor puts in only $13,000 of his own capital per wind machine and realizes a higher IRR of 18% on that capital. (This is calculated by solving equation (2) above for the IRR using the cash flows in column (5) of the table.) In other words, by "leveraging" his capital (that is, borrowing funds to finance part of the investment cost) the entrepreneur has a more profitable project. Another advantage of the leveraged project is that it takes about nine years for the NPV of the project to turn positive, that is, the project turns a profit in less time. On the other hand, by taking on the debt the entrepreneur is exposed to greater financial risk. If the project runs into trouble and the cash flow sags, the bank may foreclose on the loan in order to retrieve its capital, in which case the entrepreneur will lose his investment entirely. With a pure equity investment (that is, with no borrowing), the entrepreneur does not face this risk. In practice, the entrepreneur has a range of possible choices for the mix of equity and borrowed capital with which to finance the project. The decision

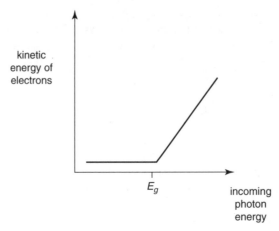

Figure 3.13. The photoelectric effect.

is influenced by the availability and price of capital of each type and the project's risks. Ultimately this is a question of risk and reward.

There is an additional complication that is important in the real world of evaluating project investments – taxes. As we saw in the gasohol and solar hot water cases, certain investments and financing alternatives will look better or worse depending on the tax consequences. The taxes paid or the tax credits earned will influence the cash flow, sometimes significantly, and this has a direct bearing on the net present value calculation that is at the heart of judging the economic viability of alternative projects.

Source: Courtesy National Renewable Energy Laboratory, Photographic Information Exchange.

Figure 3.14. Sheet of amorphous silicon solar cells produced by Iowa Thin Film Technologies.

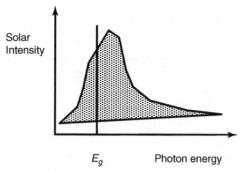

Figure 3.15. Distribution of solar intensity versus photon energy; the band gap energy, E_g, of the solar cell material is indicated.

SOLAR PHOTOVOLTAIC TECHNOLOGY

Solar photovoltaic technology is an attractive power option because it produces electricity simply and directly from a renewable source and in principle allows application on a small scale, such as an individual residence. As we shall see, however, the technology is not yet economic in the U.S. power market, and significant additional progress will be necessary before widespread deployment can occur. The technology is likely to be deployed earlier in remote applications or in less developed countries where there is no utility grid.

Photovoltaic energy is generated through the photovoltaic effect. Photons of energy $E = h\nu = hc/\lambda$ hit a semiconductor material with band gap E_g, creating an electron-hole pair. Here, as before, h is Planck's constant, c the speed of light, and ν and λ the frequency and wavelength of the light. The resulting kinetic energy of the electrons gives rise to a current. The photovoltaic effect is illustrated in Figure 3.13.

Photovoltaic solar cells are made from semiconductor materials doped to make electron donating and electron accepting regions. The materials are usually based on silicon, Si, or gallium arsenide, GaAs. A sheet of amorphous silicon solar cells is shown in Figure 3.14. The efficiency of photovoltaic materials depends upon the relationship of the band gap of the material, E_g, to the solar spectrum (see Figure 3.15). The ideal band gap is between 1.2 and 1.4 eV because this energy range is well matched to the peak intensity of the solar radiation striking the earth's surface ($\sim$1.5 eV, as noted previously.) The efficiencies that can be achieved in practice depend upon the current-voltage characteristics of the material.

For flat panel solar cells oriented perpendicular to the incoming solar radiation, the achievable efficiencies in silicon and gallium arsenide are illustrated in the following table:

Material	Ideal efficiency	Realized efficiency	E_g (eV)
Si	29%	16.5%	1.1
GaAs	36%	23.8%	1.4

Source: Courtesy National Renewable Energy Laboratory, Photographic Information Exchange.

Figure 3.16. This photovoltaic tracking system follows the sun across the sky, presenting its face perpendicular to the incoming radiation to maximize the flux of photons hitting the solar cells.

By adding engineered features to the simple flat plate solar cell the efficiency of the cell can be increased. For example, reflective backing can be added to capture a greater fraction of the incoming radiation. It is also possible to construct (at additional cost) layered materials in the solar cell; each layer is tailored to absorb a different part of the incoming solar spectrum. Photovoltaic systems can also include solar concentrators that collect solar radiation from a larger area and use mirrors to focus the incoming radiation onto the active solar cell surface. In addition to concentrators, there are tracking systems that follow the sun across the sky to maintain the favorable perpendicular orientation between incoming solar radiation and the active cell surface. One such system, located in northern California, is shown in Figure 3.16. Each of these modifications involves an engineering design decision that balances the improvement in cell performance in producing electricity versus cost.

Photovoltaic technology must be considered as a system that produces electricity in response to demand. The electricity produced in a solar cell must be conditioned to provide the required voltage and current at the time of demand. This can be straightforwardly achieved by connecting the individual solar cells together in series and parallel.

The major problem for photovoltaics, as with many other renewable energy technologies, is energy storage. Storage is especially important for photovoltaics, since electricity is only generated during daylight hours. Relying on back-up power from a conventional utility grid is a possibility, but if the utility provides a guarantee of power on demand the user of the renewable system will be charged for the cost of the stand-by source of electricity. For applications in remote or under developed areas, photovoltaics may be the only source of electricity so that intermittent supply will be considered better than no supply at all.

Effective solutions to the energy storage problem have not yet been found and this remains a key technical challenge for photovoltaics. Several alternatives are possible. Batteries are the energy storage option most frequently considered. Other forms of energy storage such as compressed air, pumped water or fly-wheels have also been investigated.

One of the most interesting photovoltaic system concepts was put forward several years ago by Texas Instruments. This concept combines photovoltaics with a fuel cell and a hydrogen bromide (HBr) circulating solution. The electricity from the photovoltaic cell reduces the HBr to hydrogen and bromine gas. The gas is stored until electricity is required, at which time the gas is sent to a fuel cell that produces electricity from the chemical energy released when hydrogen and bromine gas combine to form HBr. The system works by both the forward and backward reaction of HBr formation:

$$2HBr(aq) \rightleftharpoons H_2(g) + Br_2(g)$$

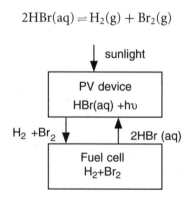

Hydrogen and bromine are challenging to handle from a safety perspective and are not ideal materials to use in a system. Their choice is dictated by the match of the electromotive force required to reduce HBr and the voltage produced by silicon solar cells. This example reminds us that not all clever schemes for making photovoltaics have yet been invented or thoroughly explored.

Photovoltaic systems must be reliable over long periods. High reliability is difficult to achieve because both oxygen and moisture must be kept from the cell material, and encapsulation integrity is difficult to assure. At present, the reliability of photovoltaic systems over a lifetime of operation is uncertain. This is not uncommon with new energy technologies, and one can expect improvement with time.

While the material purity of photovoltaic cells need not match that required for semiconductor chips, the cost of the materials is significant and the manufacturing task quite demanding. A great deal of attention has been paid to technical innovations that will lower the cost of obtaining the semiconductor materials at required levels of purity as well as to manufacturing solar cells at low cost.

Important targets for future innovation include: More efficient cell materials and cell design; integration with energy storage; and new manufacturing techniques that will lower acquisition costs.

Economics of Photovoltaics

The main contributor to the cost of photovoltaic electricity is the capital cost of the system. The capital cost is based on the peak watts that the system delivers. The reason is that the peak watt capacity of the system determines the area of solar cells that must be purchased, and this is the principal cost driver of the system. Our analysis of the capital costs of photovoltaic systems follows closely the approach presented in an early American Physical Society study on solar photovoltaic energy conversion.[3] In this analysis, the capital cost per peak kilowatt, CI ($/kw$_p$), is expressed as the sum of the power conditioning cost, the cost of the photovoltaic array, and the associated land cost. The formula is:

$$\text{CI}(\$/\text{kw}_p) = f[C_p + (C_s + C_a)/\eta q_p].$$

The variables in this equation are defined below:

   f = indirect cost factor, roughly 1.35
   C_p = power conditioning cost, about $140/kw$_p$
   C_s = site costs, about $20/m^2
   C_a = array costs, units are $/m^2
   q_p = peak solar insolation, (kw$_p$/m^2)
   η = array efficiency

To calculate the capital component of the electricity cost, K (in cents kwhr), we multiply the capital cost CI by the annual capital charge rate, ϕ, and divide by the system capacity factor, the ratio of the electrical energy actually produced by the system during the year to the energy produced if it operated continuously at peak power. If we assume a capacity factor of 0.25, the capital cost K is:

$$K(\text{¢/kwhr}) = \text{CI}(\$/\text{kw}_p) \times \phi(\text{yr}^{-1}) \times 100(\text{¢/\$})/(0.25 \times 8,766)(\text{hours/yr}).$$

Using the illustrative values for f, C_p, and C_s given previously:

$$K(\text{¢/kwhr}) = 0.062\phi[140 + (20 + C_a)/\eta q_p].$$

Next, we assume that the system is financed with a ten-year loan based on constant payments with a 10% per year interest rate. From equation (1), the annual capital charge rate, ϕ, for this loan is calculated to be 16.3%. For a peak solar flux of 1 kw$_p$/m^2, the result is:

$$K(\text{¢/kwhr}) = (0.01)[(20 + C_a)/\eta] + 1.4 = 1.4 + 0.2/\eta + 0.01(C_a'/\eta),$$

where now (C_a'/η) is the array capital cost in units of $/kw$_p$ output.

---

[3] "Solar Photovoltaic Energy Conversion." Report of a study group of the American Physical Society, chaired by H. Ehrenreich, published by the American Physical Society, January 1979.

Table 3.5. Photovoltaic cost trends and projections

	1991	1995	2000	2010–2030
Delivered electricity price (¢/kwhr)	40–75	25–50	12–20	<6
Module efficiency (%)	5–14	7–17	10–20	15–25
System cost ($/w$_p$)	10–20	7–15	3–7	1–1.50
System lifetime (years)	5–10	10–20	>20	>30
U.S. cumulative sales (MW)	75	175	400–600	>10,000

A typical cell efficiency for a system operating in the field is $\eta = 0.1$. Thus, even if the array cost were zero (i.e., $C_a = 0$), the electricity cost would be 3.4¢ per kwhr, without including any allowance for maintenance or backup power. If we include estimates for backup power of 1¢ per kwhr and maintenance costs of 0.5¢ per kwhr then the full system electricity cost L in ¢ per kwhr is:

$$L(\text{¢/kwhr}) = 0.01(C_a'/\eta) + 4.9.$$

Today, photovoltaic cells in encapsulated arrays can be manufactured for between $1 and $2 per peak watt. The following table illustrates how the electricity cost varies with the array cost:

C_a'/η ($/w$_P$)	Array cost (cents/kwhr)	System cost (cents/kwhr)
0.50	5.00	9.9
1.00	10.00	14.9
2.00	20.00	24.9

Because the cost of electricity from coal or natural gas is currently about 5¢ per kwhr or less, it is clear that photovoltaics will not penetrate the main United States market without subsidies.

Recent trends in photovoltaic costs and the current technical and economic targets of the U.S. Department of Energy and its National Photovoltaic Energy Center are summarized in Table 3.5. The table shows that although the price of photovoltaics has been falling, there is still a long way to go. The system lifetime targets have also not been demonstrated, so although the target module efficiencies are probably within reach, the projected economic costs are optimistic at best. These types of projections by advocates both inside and outside of government mean that it is difficult for the citizen, or the lawmaker, to know what to believe. The fact remains that the economic case for photovoltaics has not yet been demonstrated.

There may also be material availability problems if photovoltaics penetrate the electricity market in a big way. If 1% of electricity in the United States were generated from photovoltaics, this would correspond to an installed capacity of about 20,000 MW$_p$. The materials requirements in metric tonnes (MT) are given below if all of this electricity were satisfied from the material indicated:

Material	MT Required	Current annual production (MT)
Ge	250	76 (worldwide)
Ga	120	7 (byproduct of aluminum)
Si	120,000	10,000

These are demanding requirements, and availability and price would become important issues if large-scale market penetration occurred. Silicon has always been considered the most likely candidate for photovoltaic devices in either single crystal or amorphous form. Note also that 20,000 MW_p would require solar panel arrays (unconcentrated) occupying an area of about 2×10^8 m^2, assuming 10% conversion efficiency.

COST REDUCTIONS AND THE LEARNING CURVE

High capital costs have been a significant barrier to the market entry of wind, photo-voltaics, and other renewable energy technologies. How might the capital cost per unit of output be reduced? There are two general pathways. The first is by making innovations in the design of the system itself that will yield improvements in performance, such as increases in efficiency, availability, or system lifetime. The second is by reducing the manufacturing cost.

Here we consider the possibilities for reducing the unit manufacturing cost. We comment briefly on two ways this can be accomplished for a given system design: (1) manufacturing economies of scale, and (2) learning curves.

Economies of Scale

An important engineering question for any production process is, "What is the most efficient scale of production?" Given a technology, management system, and regulatory environment, how can production be most efficiently organized? Various combinations of inputs – capital, labor, materials – can lead to production of a quantity of output Q. For a given set of inputs $\{x_i\}$ the production function summarizes these possibilities as:

$$Q = q(x_1, x_2, \ldots x_n; \text{technology}; \text{management, etc.}).$$

We seek to organize production so that the inputs are used most efficiently. Imagine that each input is increased by a factor λ. Then the production function yields the result:

$$\lambda^\alpha Q = Q(\lambda x_1, \lambda x_2, \ldots \lambda x_n; \text{technology}; \text{management}).$$

If $\alpha > 1$, there are economies to be realized by increasing the scale of the production operation. If $\alpha < 1$, economies can be realized by decreasing the scale of the operation. At $\alpha = 1$ there are constant returns to scale. Early in the lifetime of most products there are usually increasing returns to production scale. Building larger manufacturing plants affords opportunities to allocate fixed costs of management, research and development, and design, for example, over larger numbers of units.

The Manufacturing Learning Curve

A second important aspect of manufacturing with a bearing on cost is the phenomenon of "learning." A factory with a given capital and technology base "learns" over time through the efforts of management and labor to produce the same product at progressively lower cost. As cumulative output increases, the production cost per unit declines. (This observation is generally true; the way improvements occur usually have little to do with formal classroom "learning.")

A useful rule is that each time cumulative output doubles, the cost per unit declines by a factor $f < 1$. A typical value of f is 0.85. Let $C(q)$ be the cost of the qth unit to be produced in a production run. Then we can write:

$$C(2q) = f \cdot C(q)$$

or, in general

$$C(2^n q) = f \cdot C(2^{n-1} q),$$

and

$$C(2^n) = f^n \cdot C_1$$

where C_1 is the cost of the first unit in the production run. It is straightforward to show that an alternative form of this learning curve is

$$C(q) = C_1 q^\alpha,$$

where $\alpha = -\ln f / \ln 2$. The learning curve flattens considerably as output increases.

As an example, consider a factory that produces wind machines in lot sizes of 100 at an initial cost of \$3,000 per kw rated capacity. If the learning curve has $f = 0.85$, how long must the production run be in order to drive the unit cost down to \$1,000 per kw? (Here we assume that there is no learning within each production lot, but that the effect of learning is to drive down the average cost of the wind turbines in each successive batch.) Using the learning curve formula, we see that to achieve a two-thirds reduction in cost there would have to be between six and seven doublings, which is a cumulative production of between 64 and 128 lots. The exact number of lots required according to the formula is 113, or 11,300 wind machines (see Figure 3.17.)

Advocates of renewable energy often call for government programs to "buy down" the cost of renewable energy technology. They argue that worthy new technologies are prevented from entering the market because the benefits of economies of scale or the learning curve have not been realized. A government program that acquires early production, at higher unit costs, will result in falling production costs and hence better prospects for successful market entry.

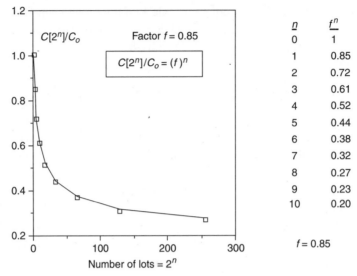

Figure 3.17. The learning curve.

Advocates of such government buy down programs have less to say about why private firms would themselves be unwilling to make investments in moving down the learning curve, if by doing so they could position themselves to achieve market penetration. Often, too, analysis of industry conditions that would favor the successful realization of economies of scale or of learning curve effects is lacking. For example, a government subsidy program that buys wind machines with the intention of achieving lower unit costs for the entire economy might be justifiable on the basis of the external benefits of the new technology, for example, reduced environmental effects. But a support program that results in creating more efficient production among a few competing producers would have a very different outcome in terms of unit production costs than a program that serves to attract a greater number of producers of small, inefficient scale.

CONCLUSION

In this chapter we have introduced several tools and concepts that are important for evaluating the economic potential of new technologies. These include: the time value of money; cash flow diagrams; the net present value and internal rate of return of a project; the effect of taxes; the role of uncertainty and sensitivity analysis; the concept of financial leverage and its implications for the financial risk borne by borrowers and lenders; and economies of scale and learning curves. These concepts have been discussed with particular reference to solar thermal, windpower and photovoltaic technologies, but they are in fact very broadly applicable. The treatment is at an elementary level. Many textbooks on engineering economics are available for readers wishing to explore these ideas in more detail.[4]

[4] See, for example, E. Paul DeGarmo, William G. Sullivan, and James A. Bontadelli, *Engineering Economy*, Macmillan, New York, 1993 (9th edition); Chan S. Park and Gunter P. Sharp-Bette, *Advanced Engineering Economics*, Wiley, New York, 1990.

APPENDIX TO CHAPTER 3

We seek an expression for the uniform annual loan payment, A, payable at the end of each year on a loan, P, received at time zero, for a loan term of N years, assuming an annual interest rate r (yr^{-1}). The cash flow diagram is

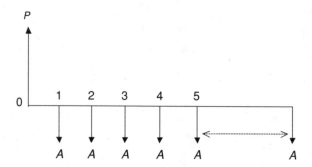

Note that this cash flow diagram has been drawn from the perspective of the borrower, who receives the loan at time zero, and makes payments on the loan at the end of each year. The initial loan is therefore drawn as an arrow above the time line, and the cash outlays to repay the loan are conversely drawn below the line. If the diagram had been drawn from the perspective of the lender, the direction of the arrows would have been reversed.

After the end of the first year, the borrower pays interest on the principal, rP, and also repays a portion of the principal, D_1, where

$$A = rP + D_1. \qquad (A\text{-}1)$$

After the end of the second year, the homeowner pays interest on the principal remaining at the end of the first year, $(P\text{-}D_1)$, and repays a further portion of the principal, D_2, where

$$A = (P - D_1)r + D_2. \qquad (A\text{-}2)$$

Substituting for D_1 in $(A\text{-}2)$ and solving for D_2 we have

$$D_2 = (A - Pr)(1 + r). \qquad (A\text{-}3)$$

After the end of the third year, the homeowner pays interest on the remaining principal, $P - D_1 - D_2$, and repays a further portion of the principal, D_3, where

$$A = (P - D_1 - D_2)r + D_3. \qquad (A\text{-}4)$$

And substituting for D_1 and D_2 in (A-4) and solving for D_3, we have

$$D_3 = (A - Pr)(1 + r)^2. \qquad (A\text{-}5)$$

Hence, by induction, we have

$$D_n = (A - Pr)(1 + r)^{n-1} \qquad (A\text{-}6)$$

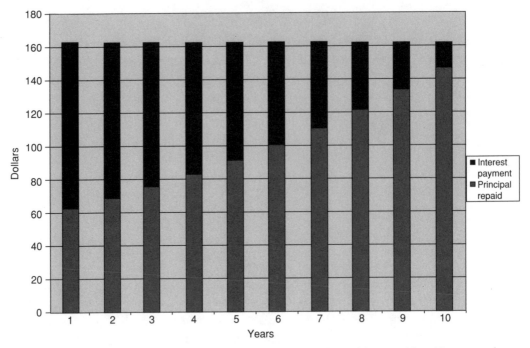

Figure 3.A.1. Interest and principal repayment for a ten-year loan of $1,000 with uniform annual payments and an interest rate of 10% per yr.

And since, by definition

$$\sum_{n=1}^{N} D_n = P,$$

we can write

$$P = (A - Pr)[1 + (1 + r) + (1 + r)^2 + \cdots + (1 + r)^{N-1}]$$

$$= (A - Pr)\left[\frac{(1 - (1 + r)^N)}{-r}\right].$$

Solving for A, we obtain

$$A = P\left[\frac{r(1 + r)^N}{(1 + r)^N - 1}\right]. \tag{A-7}$$

The term in square parentheses in equation (A-7) is called the "capital recovery factor" or the "annual capital charge rate."

In the early years of the loan, interest accounts for the lion's share of the annual payment, whereas towards the end of the term most of the payment goes toward repaying the principal. The variation of these proportions over time is shown in Figure 3.A.1 for the case of a ten-year loan of $1,000 offered at an interest rate of 10% per year.

4

Electricity from Coal

Coal is by far the most plentiful of conventional fossil fuels. The world's total coal resources have been estimated to be as much as 10,000 billion tons – enough, in principle, to meet all of the world's energy needs for 1,000 years at current rates of consumption. Coal, moreover, is widely distributed. The largest known resources are in Russia and other nations of the former Soviet Union, the United States, and China, but many other countries in every continent have sizeable deposits.

In the United States, coal is the largest energy-producing industry, accounting for nearly a third of all domestic energy production and almost a quarter of all energy consumed. The industry is a net exporter, and employs about 80,000 miners in 26 states.

The most important use of coal today in the United States and around the world is for electricity generation (see Figure 4.1). Of the billion tons of coal consumed annually in the United States, 90% is used in electric power stations, and these coal-fired plants generate more than half of the nation's electricity (see Table 4.1).

These figures make clear that coal will be an important fuel source and industry for many decades. However, there is growing awareness of the health and, especially, the environmental problems associated with its use. In this chapter we examine the technical and economic aspects of coal-fired electricity generation. This analysis provides a basis for understanding what can be done to mitigate the environmental effects of burning coal and prepares the ground for later comparisons between coal and other energy sources from an economic and environmental point of view.

COAL AS A NATURAL MATERIAL

The world's coal deposits were laid down hundreds of millions of years ago in a prolonged, complex process of compaction, chemical alteration, and metamorphosis of ancient plant materials by heat and pressure. Although one lump of coal looks very much like another, the chemical composition actually varies widely. This is largely because of differences in the conditions under which the coal was formed.

Coal has no precise chemical structure; it is composed of carbon macrocycles of extended size with some hydrogen as well as nitrogen, oxygen, and sulfur atoms. A key characteristic is the ratio of hydrogen to carbon. In most coals, the hydrogen/carbon atomic ratio is less than one. Anthracite, the hardest of all coals, contains virtually no bound hydrogen and is almost pure carbon. The amount of bound hydrogen is

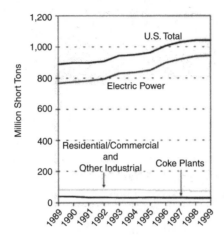

Source: Reprinted from the U.S. Energy Information
Administration, *Annual Energy Review 2000*,
http://www.eia.doe.gov/cneaf/coal/page/special/feature.pdf.

Figure 4.1. U.S. coal consumption by sector, 1989–2000.

progressively larger in bituminous, sub-bituminous, and lignite coals.[1] Other composi-
tional differences include the fraction of noncombustible "ash" (which can range from
less than 5% to more than 20% by weight), the water and sulfur content, and the trace
metal composition. These differences matter a great deal. They affect the uses to which
the coals can be put, the environmental impacts of these uses, and the costs.

Although coal does not have a unique composition or a precise chemical structure,
an empirical formula can be written for the molar proportions of its constituents. For
a typical U.S. bituminous coal this formula is:

$$C_{1.0}H_{0.8}O_{0.2}.$$

Because the atomic weights of carbon, hydrogen, and oxygen are 12, 1, and 16 respectively, this implies a carbon weight content of

$$\frac{12}{1 \times 12 + 0.8 \times 1 + 0.2 \times 16} = 75\%.$$

The variability of coal compositions is suggested by Table 4.2, which presents the compositions of two coals used extensively for electricity generation in the United States.
The much higher sulfur content of the Ohio coal is notable, and has important environmental consequences. In general, bituminous coal mined in the eastern and midwestern
states (the leading coal producers are Kentucky, West Virginia, Virginia, Pennsylvania,

---

[1] In liquid petroleum hydrocarbons, the atomic ratio of hydrogen to carbon is approximately 2 to 1, and
natural gas is mainly methane, CH_4. So, as we shall see later, making synthetic liquid and gaseous fuels
from coal generally involves adding hydrogen.

Table 4.1. U.S. electricity
generation by fuel type (year 2000)

	%
Coal	51.9
Nuclear	19.9
Natural gas	15.8
Hydroelectric	7.3
Petroleum	2.9
Renewable (geothermal, solar, wind, photovoltaic, biomass, et al.)	2.2

Source: U.S. Energy Information Administration, *Annual Energy Review,* http://www.
eia.doe.gov/emeu/aer/.

Illinois, and Indiana) is relatively high in sulfur, while western coals (mined principally in Wyoming, Montana, and North Dakota) have lower sulfur content.

The differences in composition lead to sizeable differences in the 'heating value' of coal (the energy of combustion), which ranges from 6,000–7,000 BTU per lb. for lignite to as much as 15,000 BTU per lb. for some bituminous coals. A useful expression for the heat content of coal as a function of composition is given by Dulong's formula:

$$\text{BTU/pound} = 14{,}544C + 62{,}028(H - O/8) + 4{,}050S,$$

where C, H, O, and S are the weight fractions of carbon, hydrogen, oxygen, and sulfur, respectively. The coefficients represent the approximate heating values of the various components in BTUs per lb.

Table 4.2. Composition of two U.S. steam coals

Coal components	Typical high sulfur Nobel County, Ohio (weight %)	Typical low sulfur Montana Rosebud (weight %)
Moisture	3–9	27–30
Ash	7–15	3–15
Elemental composition (basis: ash and moisture free)		
Carbon	79	72
Hydrogen	6	4
Oxygen	9	22
Nitrogen	1.5	1.1
Sulfur	4.5	0.8
Heat content	14,500 BTU/lb.	12,030 BTU/lb.

THE U.S. COAL INDUSTRY: PRODUCTION, PRODUCTIVITY, AND TRANSPORTATION

Coal production in the United States is concentrated in three regions of the country: the Appalachian region (where the most important deposits are found in West Virginia, Eastern Kentucky, and Pennsylvania); the interior region (principally Illinois, Indiana, Texas, and Western Kentucky); and the West. The Western region contains the nation's largest source of sub-bituminous, low-sulfur coal, the Powder River Basin of northeastern Wyoming and southeastern Montana. Coal production in the West is dominated by Wyoming, which currently accounts for nearly one-third of the total U.S. production. The adoption of progressively more stringent clean air regulations has increased the relative attractiveness of low-sulfur coal, and partly because of this the West has recently overtaken Appalachia as the nation's largest coal producing region (accounting for 47% in 2000, compared with 39% from the Appalachian region.)

About 60% of the coal produced in the United States (and more than 90% of Western coal) is extracted by surface or strip mining. Labor productivity levels are generally much higher in surface mines. A common measure of coal-mine labor productivity is the number of tons of coal produced per miner per hr. Labor productivity has improved significantly at both underground and surface mines in recent years, as smaller, less efficient mines have been closed and as new, larger-scale and more efficient capital equipment has been introduced. The surface mines in the West, where thick coal beds are located under thin, easily removed overburden, are by far the most productive in the country. According to the Energy Information Administration, Western surface mining, with a productivity level of about 22 short tons of coal per miner per hr in 1997, is three times as productive as its closest rival, surface mining in the Interior region, and nearly six times as productive as underground mining in Appalachia. These productivity differences are reflected in the average coal prices at surface and underground mines in the three main producing regions of the country (see Table 4.3.)

Transportation from the mine to the power plant is accomplished mainly by rail, although water and road transportation are also important in some locations. The cost of transportation accounts for a significant portion of the delivered price of coal. For low-cost western coal, much of which is shipped by rail over long distances to midwestern or eastern power plants, transportation accounts for at least 50% and, in some cases, as much as 75% of the delivered price at the power plant. For Appalachian coal, which as noted above is costlier at the mine mouth and usually is hauled over relatively short distances, the transportation cost fraction is more like 20%.

Railroad transportation of coal commonly involves dedicated unit trains – strings of about 100 hopper cars, each capable of carrying 100 tons of coal, shuttling back and forth between the mine and the power plant. One such train, carrying 10,000 tons, will deliver enough coal to supply a large 1,000 MW power plant for a single day (see in the paragraphs to follow). The coal, railroad, and power industries are strongly interdependent. The capacity and reliability of the rail infrastructure is very important to the power industry; at the same time, coal – most of it bound for power plants – is

Table 4.3. Average coal price at the mine, by region (year 2000)

Coal-producing region	Average price at the mine, 2000 ($/short ton)	
	Underground	Surface
Appalachian	26.65	24.76
Interior	22.13	16.04
Western	17.02	7.84

Source: Energy Information Administration, *Coal Home Page,* http://www.eia.doe. gov/cneaf/coal/cia/html/tbl82p01p1.html.

an extremely important cargo for the railroad industry, accounting for 44% of total tonnage hauled and more than 20% of total revenues.

Several alternatives to rail transportation have been considered. One, the so-called coal-by-wire option, entails locating the power plant at (or closer to) the mine and transporting the electricity rather than the coal, possibly requiring the construction of additional power transmission capacity. Another alternative is to transport the coal by coal slurry pipeline. A third alternative is to convert the coal to liquid or gaseous fuels at the mine mouth. With a few notable exceptions, none of these approaches has yet emerged as a cost-effective alternative to rail transport.[2]

Figure 4.2 summarizes the recent production history of the U.S. coal industry. The dramatic increase in labor productivity – more than a factor of 2 in just over a decade – reflects the shift towards larger, more efficient mines in all coal-producing regions, as well as the shift from sub-surface Eastern mines to higher-productivity Western surface mines. These changes are also reflected in the declining real price of coal at the mine during this period, and the decline in the average sulfur content of the coal. Overall coal output increased at a much slower rate during this period, and the net result was that total employment in the industry declined by nearly 50%. Many of these jobs were lost in the Appalachian region, where many older, smaller mines were closed down, causing great economic hardship in some mining communities.

CONVENTIONAL COAL-FIRED STEAM PLANTS

Most coal-fired power plants in operation today are pulverized coal units, in which finely ground coal particles are burnt in air at atmospheric pressure in a large, box-shaped boiler. The walls of the boiler are lined with heat exchanger tubes in which water is converted to steam. The steam is further heated by the hot combustion gases in a superheater and is then sent to a high pressure turbine. The rotational energy of the turbine is converted into electricity by a generator. To improve the overall energy-conversion efficiency of the plant, the steam exiting the turbine is reheated and sent to drive one or more lower pressure turbines with additional electricity generation. The

---

[2] A current example of "coal by wire" is the massive Four Corners coal fired power plant in Arizona, most of whose output is consumed in California.

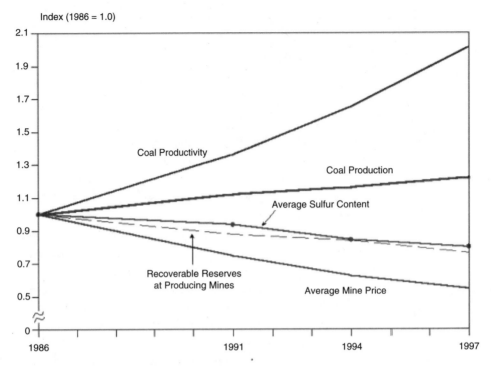

Index (1986 = 1.0)

Note: Average mine prices are indexed to constant dollars. Average sulfur content is based on coal delivered to electric utilities, reported on Form FERC-413.

Source: Reprinted from the Energy Information Administration, *Coal Industry Annual 1995*, DOE/EIA-0584(95) (Washington, D.C., October 1996), and *Coal Industry Annual 1997*, DOE/EIA-0584(97) (Washington, D.C., December 1998) Tables 1, 25, 48, 80, and 106.

Figure 4.2. U.S. coal production, productivity, prices, reserves, and sulfur content, 1986–97.

steam is cooled, condensed, and then pumped back to the boiler to begin the cycle once again.

The thermal efficiency of a power plant is important for both economic and environmental reasons, and has been a target of engineering improvements since the earliest days of the electric power industry. Modern coal plants achieve efficiencies (expressed as net electrical energy output per unit of thermal energy input) of about 35%, although older plants are usually much less efficient.[3] The efficiency of fossil plants is often expressed in terms of the heat rate, which is defined as the thermal input (in BTUs) required to produce 1 kwhr of electricity. A typical heat rate for a large modern coal-fired power plant is 9,800 BTU per kwhr. Because 1 kwhr = 3,413 BTU, this translates into an energy conversion efficiency of 3,413/9,800 = 34.8%.

[3] Coal plants in developing countries like India and China are also typically much less efficient than the most advanced coal plants in the United States and Europe. Thus, the incremental cost of improving the efficiency of coal plants is likely to be lower in many developing nations than in the United States. This has significant implications for global warming policy proposals that would cap the level of CO_2 emissions (see Chapter 6).

The operating temperature is a key design consideration in coal plants. The higher the steam temperature at the inlet to the turbine, the greater the energy conversion efficiency.[4] This lowers fuel input costs per unit of electrical output and may reduce plant capital costs too.

A key constraint on steam temperature is imposed by corrosion in the turbine. At higher temperatures the rate of corrosion increases. This in turn tends to result in increased operating and maintenance costs and reduced plant lifetimes. Higher temperatures in the boiler also result in higher production rates of nitrogen oxides (NO_x), as well as more ash melting and fouling of the heat exchanger tubes, which reduces heat transfer efficiency and plant reliability.

Direct burning of coal produces large quantities of solid, liquid, and gaseous effluents, including solid coal ash, waste water, atmospheric emissions of sulfur oxides, nitrogen oxides, airborne particulates from the ash, arsenic and mercury (which are frequently present in trace amounts in the coal), and, of course, carbon dioxide. Each of these poses an environmental problem that must be addressed.

Roughly 70% of the coal ash is carried off in the flue gases, from which it must be recovered by passing the gases through electrostatic precipitators or large bag filters. These filtration mechanisms are less efficient at removing the very small ash particles, which are the most troublesome for human health. Although they only constitute a small fraction of the weight of the ash, these "fines" account for most of the surface area, making them particularly effective absorbers of sulfur oxides and the trace metal by-products from combustion. And because they are less likely to be filtered out on their way to people's lungs, they are also more biologically potent.

Since the late 1970s regulations have limited the amount of sulfur that can be emitted by power plants. As a result, many coal plants today are equipped with flue gas desulfurization (FGD) systems that remove sulfur dioxide (SO_2). In most of these units, the flue gases are contacted with a slurry of limestone in a spray tower. The SO_2 reacts with the limestone in these "wet scrubbers" to produce a hydrated calcium sulfite or sulfate (gypsum). The resulting sludge is produced in large volumes and must be disposed of. The scrubbers also reduce the overall thermal efficiency of the plant. Typically 3% to 8% of the energy output is used to run the pumps and fans and to reheat the flue gas so as to prevent corrosive condensation in the stack.

In many plants, especially those using high-sulfur coal, the coal is washed before combustion to remove the "free" sulfur – sulfur that is not bound chemically into the carbon skeleton but present as metal sulfide inclusions such as iron sulfide (Fe_2S_3).

---

[4] The theoretical maximum thermodynamic efficiency of a Rankine steam cycle plant (the Carnot efficiency) is given by:

$$\frac{T_1 - T_2}{T_1},$$

where T_1 is the maximum steam temperature and T_2 is the ambient temperature (in degrees Kelvin). Thus, the plant's overall efficiency can be increased by using higher superheat conditions in the steam cycle.

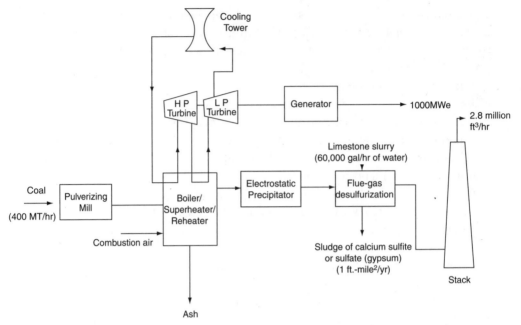

Figure 4.3. Conventional pulverized coal power plant schematic (basis: 1000 Mwe).

Coal washing helps to reduce the gaseous sulfur emissions, though it also produces large amounts of waste water and sludge.

Modern coal plants are also equipped with treatment systems to reduce the emissions of nitrogen oxides. In the following chapter, we will describe in more detail the environmental problems caused by sulfur and nitrogen oxide emissions from power plants and the strategies that are available to mitigate them. For the moment, it is enough to say that pollution control systems account for a large fraction of the total cost of generating electricity from coal – as much as 40% of the capital cost and 35% of the operating cost according to one estimate.[5]

A schematic process flow diagram for a conventional coal plant is shown in Figure 4.3.

Material Balance Considerations

A typical large coal plant produces 1,000 MWe of electric power (equivalently, 1 million kwhr of electricity per hr.) Assuming a plant heat rate of 9,800 BTU per kwhr and a heating value of 25 million BTU per ton (typical of bituminous coals), the plant consumes:

$$10^6(\text{kwhr/hr}) \times 9,800(\text{BTU/kwhr}) \times (1/25 \times 10^6)(\text{tons/BTU})$$
$$\approx 400 \text{ tons of coal per hour}$$

---

[5] Richard E. Balzhiser and Kurt E. Yeager, "Coal-fired power plants for the future," *Scientific American*, p. 92, (1987).

or about 10,000 tons per day. Such a plant will discharge nearly 3 million cubic feet of flue gas per hour through the stack. As already noted, the main pollutants in the flue gas are sulfur dioxide (SO_2), nitrogen oxides (NO_x), particulates from the combustion process, trace elements, and carbon dioxide.

For a coal containing 80% carbon, 8% ash, and 1% sulfur by weight (a fairly high quality coal), about 800 tons of ash will be produced per day, along with 8,000 tons of carbon and 100 tons of sulfur in the stack gas. Because the atomic weights of carbon, sulfur, and oxygen are 12, 32, and 16, respectively, this is equivalent to $8,000 \times (44/12) \approx 30,000$ tons per day of carbon dioxide and $100 \times (64/32) = 200$ tons per day of SO_2.

ECONOMIC ANALYSIS: THE LEVELIZED COST OF ELECTRICITY

In this section we develop an expression for the lifetime levelized cost of electricity from a coal-fired power plant. This can be thought of as the constant price which, if received by the plant owner for each kilowatt hour of electricity generated by the plant over its lifetime, will provide a flow of revenues just sufficient to cover all of the costs incurred in building and operating the plant throughout its life. The lifetime levelized cost is a simple and useful indicator of economic merit. It can be used to make economic comparisons of alternative sources of electricity with quite different cost characteristics. Other things being equal, the alternative with the lowest levelized cost will be preferred.

The costs incurred by the coal plant owner can be divided into three categories:

A. Capital cost
B. Fuel cost
C. Operating and maintenance (O&M) cost

We next calculate the corresponding levelized cost component for each of these categories.

Levelized Capital Charge

The 'levelized capital charge' is the component of the levelized electricity cost that yields a stream of revenues just sufficient to cover the cost of plant construction. To calculate it, we first make the simplifying assumption that the number of kilowatt hours of electricity the plant produces each year remains constant. Let this annual output be H kwhr. If I_o is the plant construction cost, we might then suppose that the levelized capital charge would be just I_o/HL, where L is the plant operating life (in years).

To see why this is incorrect, we need only consider the present worth (i.e., the equivalent value at the beginning of plant life) of the revenues that would be received by applying such a charge. The situation is shown in the cash flow diagram below, where we assume for simplicity that the revenues are received not at the time that each kilowatt hour is produced but rather in a lump sum, I_o/L, at the end of each year.

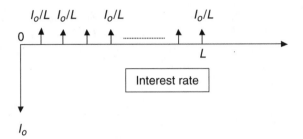

Clearly, the sum of the present worths of this series of annual cash flows:

$$\frac{I_0/L}{(1+i)} + \frac{I_0/L}{(1+i)^2} + \frac{I_0/L}{(1+i)^3} + \cdots + \frac{I_0/L}{(1+i)^L}$$

is less than the initial investment cost I_0.

The actual levelized capital charge, e_c, can be obtained by solving the present worth balance:

$$I_0 = \sum_{n=1}^{L} \frac{e_c H}{(1+i)^n}.$$

From this we have that the levelized capital charge,

$$e_c = \frac{(I_0/H)}{\displaystyle\sum_{n=1}^{L} \frac{1}{(1+i)^n}} = \frac{i I_0/H}{\left[1 - \dfrac{1}{(1+i)^L}\right]}. \tag{1}$$

Note that in the limit as the interest rate i approaches zero, the formula predicts that the levelized capital charge approaches I_0/HL, the simple undiscounted ratio of initial capital cost to lifetime kilowatt hours produced, as we would expect.

A representative initial investment cost for a new 1,000 MWe coal plant is 1.5 billion dollars (i.e., \$1,500 per kw). Assuming a plant lifetime of 30 years, a plant capacity factor of 80% (equivalent to 7,000 hours of full power operation) and an interest rate of 10% per year, the levelized capital charge is 2.3¢ per kwhr.

The levelized capital charge is strongly affected by the value of the interest rate.[6] Equation (1) above can be rewritten as

$$\frac{e_c}{e_c^o} = \left[\frac{i L}{1 - \dfrac{1}{(1+i)^L}}\right],$$

where $e_c^o (= I_0/HL)$ is the undiscounted capital charge (i.e., when $i = 0$). Figure 4.4

---

[6] Note that power plant projects involve a combination of equity capital and loaned funds. As we saw in Chapter 3, lenders (banks or bondholders) are usually not prepared to extend loans to cover the full cost of construction. They require a certain fraction of equity in the project, so that the plant owners share in the risks. Thus, what we have referred to here as the interest rate is in fact a blend of the interest rate on borrowed capital and the expected rate of return on private equity investment. Also, the effects of taxes are not considered in this simplified analysis.

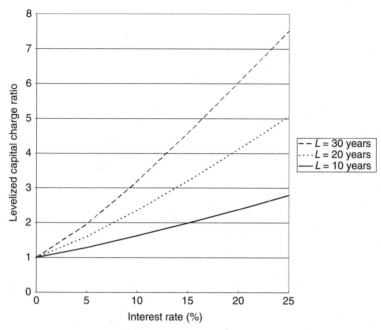

Figure 4.4. The levelized capital charge (as a multiple of the undiscounted capital charge) versus interest rate.

shows the dependence of the levelized capital charge (expressed as the ratio e_c/e_c^o) on the interest rate, for three different values of the plant operating lifetime. As can be seen, higher interest rates result in higher levelized capital charges. The reason is clear: As the interest rate increases, the discounted value of the future revenue streams declines. So, the uniform charge required to produce revenues equivalent in a present worth sense to the initial investment cost must rise.

Levelized Fuel Cost

We next calculate the fuel cost component of the levelized electricity cost. The average price of coal delivered to U.S. electric utilities in the year 2000 is shown in Table 4.4.

Let us assume a delivered price of coal at the power plant of $40 per ton (i.e., including transportation costs). Let us also assume a plant thermal efficiency of 35% and a coal heating value of 13,000 BTU per lb. This leads to a fuel cost in cents per kilowatt hour of:

$$40(\$/ton) \times 1/2,000 \text{ (tons/lb.)} \times 1/13,000 \text{ (lb./BTU)} \times 1/0.35 \text{ (thermal/electric)}$$
$$\times 3,413 \text{ (BTU/kwhr)} \times 100(\text{¢}/\$) = 1.5\text{¢/kwhr.}$$

The market price of coal is unlikely to remain unchanged over the thirty-year lifetime of the plant, and the calculation of lifetime levelized cost must incorporate an assumption about the future fuel price trend. In the Appendix to this chapter, we derive

Table 4.4. Average price of coal delivered to U.S.
electric utilities, by region (year 2000)

	Nominal dollars per short ton
New England	40.16
Middle Atlantic	31.16
East North Central	26.35
West North Central	14.69
South Atlantic	34.81
East South Central	27.28
West South Central	19.08
Mountain	21.13
Pacific	23.09

Source: Energy Information Administration, *Coal Home Page*,
http://www.eia.doe.gov/cneaf/coal/cia/html/t92p01p1.html.

an expression for the lifetime levelized fuel charge for the special case of an exponential rate of increase (or decrease) in the fuel price. The equation is as follows:

$$e_F = e_F^o \frac{r}{r-x} \left[\frac{1 - e^{-(r-x)L}}{1 - e^{-rL}} \right] \tag{2}$$

where e_F is the levelized fuel charge, e_F^o is the fuel price at the beginning of the plant life, x is the annual rate of increase in the fuel price, and r is the interest rate, where in this expression interest is assumed to be compounded continuously (see the Appendix for a more extensive discussion of the frequency of compounding.)

At an interest rate of 10% per year, an initial coal price of 1.5¢ per kwhr, and a coal price escalation rate of 4% per year, the lifetime levelized fuel cost would be 2.2¢ per kwhr.

Levelized Operating and Maintenance Cost

The lifetime levelized O&M cost component can be calculated using an equation analogous to equation (2) above. Operating and maintenance expenses at new coal plants today are typically on the order of 0.5¢ per kwhr. Again assuming an interest rate of 10% and an escalation rate for O&M expenses of 4%, the O&M cost levelized over the lifetime of the plant would be 0.73¢ per kwhr.

With these assumptions, the levelized cost of electricity from the coal plant is thus $2.3 + 2.2 + 0.73 = 5.2$¢ per kwhr.

The lifetime levelized cost of electricity is a simple and useful indicator of economic merit. The method used to calculate it here is quite general and can be used to make economic comparisons of alternative sources of electricity with quite different capital, operating, and fuel cost characteristics. Other things being equal, the alternative with the lowest levelized cost will be preferred. It is important to keep in mind, however, that the levelized cost calculation does not indicate whether an investment in any particular technology is actually justifiable economically. The answer to that question also depends on the price that the plant owner expects to receive for the electricity over the plant lifetime. As the deregulation of the electric power industry proceeds, the price of electricity is increasingly being determined in a competitive marketplace by the balance of supply and demand.

ADVANCED COAL TECHNOLOGIES

Increasing the thermal efficiency of coal-fired power plants has both economic and environmental benefits. Higher thermal efficiencies mean lower fuel costs and probably also lower capital costs per unit of electrical output. Efficiency improvements also result in lower atmospheric emissions of sulfur oxide, nitrogen oxides, and carbon dioxide per unit of electrical output. A long series of technological innovations has helped to raise the thermal efficiency of coal plants from about 5% at the beginning of the century to about 35% today. New coal technologies promising higher efficiencies and lower emissions are now available or under development, including atmospheric and pressurized fluidized bed combustion systems and integrated coal gasification/combined cycle combustion technologies (IGCC).

Atmospheric Fluidized Bed Combustion (AFBC)

In atmospheric fluidized bed combustion units, coal and limestone are crushed and fed to a bed of unburnt coal, ash, and inerts, which is fluidized by air injection. The air also is the combustion agent for the coal. The turbulent mixing of air and coal in the bed facilitates very efficient combustion at relatively low temperatures (800–900°C) – about half the temperature in a conventional boiler.

Fluidized bed combustion has several advantages relative to conventional pulverized coal boilers:

- At the lower combustion temperatures in the fluidized bed, there is less production of NO_x from reactions between atmospheric nitrogen and oxygen (though nitrogen in the coal itself may still reach the temperature required to form NO_x). Also, the lower vapor pressure of metal sulfates and chlorides means that these constituents are more likely to remain in the bed itself than to be released in the flue gas.
- Desulfurization is achieved at much lower capital cost by adding crushed limestone to the bed to absorb the sulfur oxide; the sulfur is removed in dry form (as calcium sulfite).
- More efficient heat transfer is achieved in the fluidized bed through direct contact between the fluidized particles and steam tubes which can be buried in the bed; heat conduction is more efficient than radiant or convective heat transfer.
- The fluidized bed is less sensitive to variations in coal quality. Higher ash content is less of a problem, and the bed can operate with lower quality coal.

A significant operational problem with atmospheric fluidized bed combustion, how-ever, is the severe abrasion and corrosion of internal parts in the bed itself.

Pressurized Fluidized Bed Combustion (PBFC)

Like conventional pulverized coal units, atmospheric fluidized bed combustion systems are based on the Rankine steam cycle. The combustion gases are used to raise steam, which in turn drives a turbine. There are no significant thermodynamic efficiency gains. In pressurized fluidized bed systems, the hot pressurized combustion gases are run directly through a gas turbine. The exhaust gases are then used in a secondary cycle to generate steam which in turn drives a steam turbine. The efficiency of this combined cycle system is several percentage points higher than the simple steam cycle plants, which offers the prospect of significant cost savings.

The maximum operating temperature at the inlet to the gas turbine is 1250°C, a limit imposed by turbine materials considerations. But the plant cannot operate at this temperature because the temperature in the fluidized bed itself must not exceed the melting point of coal ash (i.e., 850–900°C). The effect is to reduce the thermodynamic efficiency below what it would otherwise have been. An offsetting benefit of the lower combustion temperature is that there is less NO_x formation and greater removal of sulfur.

Integrated Gasification Combined Cycle System (IGCC)

The integrated coal gasification combined cycle system (IGCC) gets around the problem of low turbine inlet temperature by first gasifying the coal and then, in a second stage, burning the fuel gas at high temperature in a gas turbine. As with the PBFC, the hot exhaust gas from the gas turbine is used to raise steam in a secondary cycle. Reducing conditions in the coal gasifier along with the relatively low temperature allow for a high level of removal of sulfur, nitrogen, and chlorine, while the separation of the gasification and combustion stages allows the high temperature capabilities of the gas turbine to be fully exploited (although at these higher temperatures the production of NO_x from atmospheric nitrogen is increased). Coal gasification technology is described in more detail in Chapter 12 of this book.

APPENDIX TO CHAPTER 4

Continuous Compounding of Interest

In the present value calculations in this chapter and Chapter 3 we have assumed peri-odic compounding of interest with cash flows occurring at the end of each compound-ing period. Implicitly, in fact, we have been assuming annual payments and annual compounding.

In practice, of course, payments don't always occur annually, and interest is compounded over many different intervals. Interest on bank deposits is often computed and paid quarterly, for example. In such cases it is common to see reference to an annual rate of interest, compounded quarterly. For example, a financial service firm might offer a loan at 'an annual interest rate of 12%, compounded quarterly.' What this really means is that the interest per period (in this case, per quarter) is $12/4 = 3\%$, and that the effective annual interest rate is

$$(1 + 0.03)^4 - 1 = 1.1255 - 1 = 0.1255, \text{ that is, } 12.55\%.$$

The example shows that when compounding is done more frequently than annually, it is important to differentiate between the effective annual rate and the nominal rate (often also referred to as the "annual percentage rate"). In general, if i is the interest rate per period and m is the number of compounding periods per year:

- the effective annual rate of interest, $i_a = (1 + i)^m - 1$;
- the nominal rate of interest, $r = i \cdot m$;

and $i_a = (1 + r/m)^m - 1$.

In the limiting case of continuous compounding:

$$i_a = \lim_{m \to \infty} \left(1 + \frac{r}{m}\right)^m - 1,$$

and since

$$i = r/m,$$

we have

$$i_a = \lim_{i \to 0}(1 + i)^{\frac{r}{i}} - 1$$
$$= e^r - 1$$

or

$$r = \ln(1 + i_a).$$

So, for example, for a nominal interest rate of 10% with continuous compounding, the effective annual rate $= 10.52\%$

Calculating the Levelized Cost for the Case of Continuous Cash Flows

In the main chapter we derived the levelized capital charge, e_c (in cents per kwhr), such that if it were applied to each kilowatt hour of electricity produced and sold over the life of the power plant, the present worth of the resulting revenues would be equal to the initial investment cost of the plant. We assumed in the derivation that the plant

produced the same number of kilowatt hours each year, and that the revenues were
collected at the end of each year.

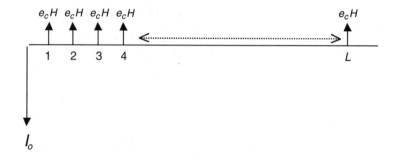

Similarly, if we have a series of variable cash flows, A_j, occurring at discrete points in
time, we can define an "equivalent levelized" cash flow, A_L such that the present worth
of a uniform series of these cash flows is equal to the present worth of the actual series:

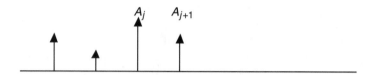

Equivalence in a present value sense requires that

$$\sum_{j=1}^{L} A_L(1+i)^{-j} = \sum_{j=1}^{L} A_j(1+i)^{-j}. \tag{A.1}$$

In the case of cash flows occurring continuously over time (a closer approximation
to reality for power plant operators who continuously receive revenues from customers
and continuously make payments to fuel suppliers, employees, and others), we can
similarly define a levelized cash flow rate, $\bar{A}_L$, which is equivalent in a present worth
sense to an actual time-varying cash flow rate $\bar{A}(t)$:

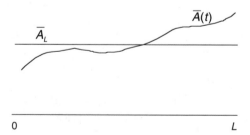

We can write, by analogy with equation (A.1) previously

$$\bar{A}_L \int_0^L e^{-rt}dt = \int_0^L \bar{A}(t)e^{-rt}dt, \tag{A.2}$$

where in this expression r is the nominal interest rate for continuous compounding discussed in the previous section.

For the special case of exponential growth of actual outlays

$$\bar{A}(t) = A_0 e^{xt},$$

The levelized cash flow rate in this case is given by

$$\bar{A}_L = \frac{\int_0^L A_0 e^{-(r-x)t} dt}{\int_0^L e^{-rt} dt} = A_0 \frac{r}{r-x} \left[\frac{1 - e^{-(r-x)L}}{1 - e^{-rT}} \right].$$

5

Controlling Acid Rain from Coal-fired Power Plants

The acid rain case considered in this chapter is an example of good government decision making. The history of federal acid rain control legislation demonstrates that it is possible for the government to arrive at an environmental control strategy that minimizes the costs of bringing about reductions, and that is at least consistent with analyses of the costs and benefits of alternative courses of action. The rational approach to decision making in this case is very different from the story of the federal gasohol program presented in Chapter 2.

The public is greatly concerned about the environmental impact of emissions from coal-fired electric power plants. An important question is how much of society's resources should be spent on reducing the environmental impact of electricity generated from coal. As discussed in the preceding chapter, there are many different consequential emissions that must be considered, including carbon dioxide (CO_2), sulfur dioxide (SO_2), oxides of nitrogen (NO_x), particulates, heat, and solid and liquid wastes.

In this chapter we consider the gaseous emissions of SO_2 and NO_x. These emissions form acids when combined with moisture in the atmosphere. The possible result is the phenomenon of "acid rain," in which rain falling at a considerable distance from the originating plant (perhaps across a national border) has high acidity. This high acidity rain can harm forests, vegetation, lakes, and the fish the lakes contain; indeed, acid rain impacts the entire ecology. Concern over acid deposition from power plant emissions of SO_2 and NO_x led to an important series of federal legislative initiatives including, most recently, the 1990 Clean Air Act Amendments, one of the most significant pieces of environmental legislation of recent years. The effect of increasingly stringent government regulation has been to bring about significant reductions in total emissions of SO_2 and NO_x, even as the economy has continued to grow. Between 1980 and 1998, for example, power plant emissions of SO_2 and NO_x declined by 24% and 13%, respectively, despite an increase of almost 60% in electricity generation during this period.[1]

The process of wet (through rain) and dry (carried by particulates) acid rain is described in Figure 5.1. The tall smokestack, whose purpose is to disperse or transport undesirable emissions away from the plant location, is the cause of the problem. The motivation for dispersal is sensible: To dilute noxious or undesirable emissions to the extent that they no longer cause any harm when they fall to the ground. The motivation

---

[1] See U.S. Environmental Protection Agency, *National Air Pollutant Emission Trends: 1900–1998*, Chapter 3, at http://www.epa.gov/ttn/chief/trends/trends98/. In 1998, power plant emissions of SO_x and NO_x were 13.2 million short tons and 6.1 million short tons, respectively.

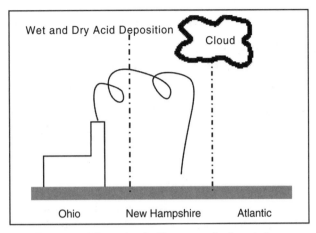

Figure 5.1. Schematic of utility atmospheric emissions.

for transport is more questionable: To transport the noxious emissions to a location where their deposition doesn't matter. In the cartoon representation of Figure 5.1, the plant owner intends its atmospheric emissions to be transported to the Atlantic Ocean, where the acid deposition can do no harm, but the main portion of the emissions actually falls on intermediate locations, such as the forests of New Hampshire or Canada.

The smokestack, a technological advance of the nineteenth century, is the cause of to-day's environmental conflict. The effect of the smokestack is to separate the source of the emission from the location that bears the burden it imposes. If there was no smokestack, the emissions would fall around the plant site, and there would be a greater incentive to pay the costs of mitigation. The utility and its customers would be confronted with the need to balance the benefits of electrical generation against the costs imposed by the environmental pollution. Because of the source's (and the customer's) proximity to the environmental impact, the impact would be more likely to be considered and mitigated. But the existence of the smokestack means that the utility and its customers in the Midwest benefit from the electricity, while the environmental costs are imposed on New England or Canada. It is a classic externality. Unless corrective action is taken, the utility will emit more pollution than is desirable. But how much mitigation should there be, and who should pay for it? Clearly some sort of mechanism is required to balance the benefits with the costs and to allocate the costs among the various affected parties. Such a mechanism should be based on the best available technical and economic information about the severity of the environmental impacts and the costs of mitigating them.

The Mechanisms of Acid Rain

The phenomenon of acid rain involves a series of steps: (1) emission of pollutants; (2) atmospheric chemical transformation; (3) long range transport and deposition; (4) environmental effects. In order to understand the nature of the environmental problems presented by acid rain, each of these steps must be described.

Emission of Pollutants

Much is known about the sources and quantities of acid rain precursor pollutants. In the United States, the utility sector is the principal source of SO_2 emissions and also a major source of NO_x, as shown below:

Source	SO_2	NO_x
Utility	65%	30%
Industry	25%	25%
Transportation	3%	40%

Almost all of the utility sulfur dioxide emissions come from coal-burning plants. Moreover, a relatively small number of plants account for the bulk of the pollution. These can be identified by their size, the fact that they burn higher sulfur coals, or because they do not have scrubbers to remove SO_2 from the stack gas. In contrast, the industrial sources (except for a few large installations) and the transportation sources (mostly private cars) are far more numerous. The utility sector is, therefore, a prime target for regulation.

Atmospheric Chemical Reactions

The combustion of sulfur during the burning of coal yields sulfur dioxide. This is oxidized in the atmosphere to SO_3, which in turn forms sulfuric acid when it comes into contact with moisture in the air (the mechanism of the hydration of SO_3 to sulfuric acid is not well understood):

$$2SO_2 + O_2 \rightarrow 2SO_3$$

$$2SO_3 + H_2O \rightarrow H_2SO_4(aq)$$

$$H_2SO_4(aq) \rightarrow 2H^+ + SO_4^{2-}.$$

For nitrogen oxide emissions, the chemical transformation is similar. However, the mechanism is quite complicated, and photochemical reactions are also involved.

$$NO_x \rightarrow HNO_3(aq) \rightarrow H^+ + NO_3^-.$$

Acidity is measured in terms of the concentration of the hydrogen ion H^+. The conventional unit of measurement is pH, which is defined as:

$$pH = -\log_{10}[H^+]$$

where $[H^+]$ is the hydrogen ion concentration expressed in moles/liter. In pure water, the equilibrium dissociation

$$H_2O \rightleftharpoons H^+ + OH^-$$

leads to a pH of 7 at 25°C and 1 atmosphere pressure. The presence of positively charged cations or negatively charged anions that can bind OH^- and H^+ in the presence of the water dissociation will influence the equilibrium amount of H^+.

In order to reach a conclusion about whether emissions are acidifying natural waters, it is necessary to define the state of "natural acidity." The conventional definition of natural acidity is the pH that occurs when natural water is saturated with CO_2 from the atmosphere according to the equilibrium:

$$CO_2(g) + H_2O \rightleftharpoons H_2CO_3(aq).$$

The resulting pH is 5.6, which corresponds to a hydrogen ion concentration, $[H^+]$, of 2.5 micromoles/liter. Of course, other natural compounds may be present in fresh water that can influence its pH. The range encountered is $4.9 < pH < 6.5$. Most fresh water is slightly acidic, that is, $pH < 7$.

Long Range Transport and Deposition

Sophisticated atmospheric transport models can predict the transport of material emitted at a plant location in, say, the Midwest to distances of many thousands of kilometers. These models include the random effect of wind strength and direction (prevailing easterly), as well as the dispersion expected from diffusion effects. Efforts have been made to calibrate the model predictions with field measurements. One interesting proposal would add a sulfur radioisotope tag to the emissions from a power plant and measure its deposition at large distances from the plant for the purpose of developing more accurate models.

There are two principal deficiencies in the existing models. First, the sulfur and nitrogen pollutants are not transported entirely in gaseous form. Some are emitted or transported on small solid particles. This not only affects the pollution transport, but it also has an important influence on the chemistry, because it introduces the possibility of heterogeneous catalytic transformation. Second, deposition (wet and dry) involves not only the flux of atmospheric pollutants onto the earth's surface but also the nature of the surface, for example, open ground versus ground covered by vegetation.

Despite these gaps in knowledge, surprisingly simple assumptions lead to useful results. For example, if one simply assumes that "what goes up, must come down," it follows that a fraction of the material emitted will fall on the land area of interest. No statement can be made about the distribution within this area, and the remaining material is ignored (it falls out to sea, as indicated by the cartoon in Figure 5.1). The policy implications of this simple model will be addressed in the next section. But it is important to understand that the central physics of all problems cannot be grasped so simply. Indeed, the case of CO_2 and global warming discussed in Chapter 6 is an excellent example of a problem in which simple transport rules are not applicable.

Environmental Effects of Acid Rain

In contrast to what is known about the steps leading to acid rain deposition, there is little agreement about the environmental effects of acid rain and almost no information about the time it might take for an ecosystem damaged by acid rain to return to its "natural" state.

Acid deposition influences soils, forests, agricultural crops, lakes and streams, and materials and buildings. There is general agreement that acid rain hurts trees. There are dreadful examples of damage to forests in Germany and eastern Europe. The effect of acid deposition on soils and crops depends upon the acidification of the soil; some have argued that acid deposition may actually improve crop yields in certain circumstances. The damage suffered by materials and buildings occurs mostly in urban areas and is attributable to localized automobile emissions. The policies proposed to regulate utility emissions will not change this situation.

The main source of controversy is the effect of acid rain on lakes and streams and on the fish and other wildlife that live in or near these waters. No reliable inventory exists for the number and location of lakes and streams that have suffered increasing acidity from anthropogenic emissions. There is little long-term data (an exception is the data from Hubbard Brook, New York) that documents historical increase in acidity and correlates such an increase with adverse affects on fish. Undeniably certain lakes in New York, New Hampshire, Massachusetts, and elsewhere along the eastern seaboard, as well as in Canada, have experienced enormous increases in acidity that can be attributed only to acid rain, and fish populations in these lakes have declined. But no quantitative estimates exist of the extent of this damage, and it is not clear whether the environmental benefit resulting from the expensive regulatory policies that have been considered is worth the cost. In the case of acid rain, David Stockman, Director of the Office of Management and Budget under President Reagan, made the flip observation that acid rain regulations would have us paying $75 per lb. of fish saved.

It is worth pointing out that the levels of atmospheric pollution from SO_2 and NO_x under discussion do not adversely affect human health. If existing pollution levels measurably impaired human health, the acid rain problem would presumably have received attention more promptly, and mandated reductions in emissions would have been greater.

Mitigation Strategies

There are several options for reducing the emissions that produce acid rain. For SO_2 emissions the options include:

- Installing scrubbers to remove SO_2 from the flue gas;
- Washing the coal more thoroughly prior to combustion to remove free sulfur;
- Introducing more efficient coal combustion technology;
- Using lower sulfur coal, or switching to natural gas.

Other strategies that will reduce the environmental impacts of acid rain include:

- Reducing energy consumption by encouraging conservation and greater end-use efficiency;
- Directly mitigating the effects of acid deposition, for example, by liming lakes.

Several of these options would obviously be effective in reducing NO_x emissions too. In the remainder of this chapter we will focus mainly on SO_2 emission controls.

Many coal plants today are equipped with wet scrubbers, in which the SO_2 in the flue gas reacts with a slurry of limestone or lime in a spray tower. The chemical reactions for the scrubbing process are:

$$SO_2 + H_2O \rightarrow H_2SO_3$$

$$\underset{\text{limestone}}{CaCO_3} + H_2SO_3 \rightarrow \underset{\text{gypsum}}{CaSO_4} + CO_2$$

$$\underset{\text{lime}}{Ca(OH)_2} + H_2SO_3 \rightarrow CaSO_4 + H_2O$$

Since the scrubbing process removes sulfur stoichiometrically, large amounts of limestone and lime are required, and large amounts of gypsum are produced. The scrubbing process transforms the pollutant from gaseous form to the more convenient solid form. It is still necessary to dispose of the gypsum, which fortunately has reuse value in building materials.

Scrubbers cost about $600 per kw of generating capacity and represent a major investment for the utility. As discussed in the previous chapter, scrubber operations also consume several percent of the energy output of the power plant.

A Model to Assess Alternative Mitigation Strategies

Beginning in the 1970s, many legislative proposals were put forward to reduce the level of emissions from utilities and other coal-burning industries. Members of Congress with an interest in acid rain control legislation were aware of the need to balance mandated reduction levels against the cost of achieving them. The quest for balance between the benefits and costs of reduction was prompted not only by a sense of fairness, but also by the political reality that some regions of the country, notably the Midwest, would bear most of the increase in electricity production costs, while the benefit of reduced acid rain would be mainly experienced in the East.

In order to assess the relative virtues of alternative proposals, it is necessary to have a model that quantitatively relates source emissions (by type, location, and time) to deposition (by chemical type, location, and time). Constructing such a model (or sets of models) and verifying the predictions is a very large task. The government established an interagency program called the National Acid Precipitation Assessment Program (NAPAP) to develop such a model and to gather information about environmental damage from acid rain. Here we rely on a very simple linear model that aggregates many of the important features that must be addressed (including regional variations, seasonal and diurnal variations, weather conditions, wet and dry deposition, and surface boundary conditions) in order to arrive at a simple relationship between reductions in rates of emissions and resulting rates of deposition. (The model is based on the work of Professor Jay Fay and Dr. Dan Golomb at MIT in the early 1980s.)

The model assumptions are: (i) the ratio of deposition to emissions is about 1:5 for both sulfur and nitrogen, where the deposition region is the northeastern United States; and (ii) the concentration of hydrogen ion deposited is obtained by assuming the amount deposited annually is dissolved in the volume of annual rainfall.

Thus, the formula for the hydrogen ion concentration in the rainfall is:

$$[H^+] = (1/5)[2E(SO_2) + E(NO_x)]/V + \text{``natural''},$$

where $E(SO_2)$ is the annual emitted amount of SO_2, $E(NO_x)$ is the annual emitted amount of NO_x, and V is the volume of annual rainfall. The factor 2 in the formula is present because each molecule of SO_2 is assumed to form a sulfate ion that is associated with 2 H^+ ions. The quantity "natural" is the assumed natural background hydrogen ion concentration. In 1979, the values of these various quantities were:

$$E(SO_2) = 2.2 \times 10^{11} \text{ moles}$$
$$E(NO_x) = 1.8 \times 10^{11} \text{ moles}$$
$$V = 1.6 \times 10^{15} \text{ liters},$$

and the natural background hydrogen ion concentration can be taken as 2.5×10^{-6} moles/liter. For these values, the incremental concentration of SO_2 in the rainfall was $2 \times (2.2 \times 10^{11})/(5 \times 1.6 \times 10^{15}) = 55 \times 10^{-6}$ moles/liter, and for NO_x it was $1 \times (1.8 \times 10^{11})/(5 \times 1.6 \times 10^{15}) = 22 \times 10^{-6}$ moles/liter.

We next construct a formula based on the assumptions of the simple model for the hydrogen ion concentration that predicts the result of a decrease in emissions below the 1979 levels. Using the values given above, we arrive at the formula:

$$[H^+] = (10^{-6})\{55[(100 - x)/100] + 22[(100 - y)/100] + 2.5\}\text{moles/liter},$$

where x is the percentage reduction from 1979 levels for SO_2 and y is the percentage reduction of NO_x.

In practice the acidity of the rainfall may be higher than the acidity in the impacted lakes and streams because equilibrium may not be attained between the rainfall and the lakes, and because other neutralization processes occur in the soil, lakes, and streams owing to the presence of other chemical species. Nevertheless, the formula is useful for estimating the results of various legislative proposals.

Alternative Cases of Acid Rain Reduction

Simple models such as the box model described above can be used to estimate the change in pH from the base 1979 case that would result from various proposals to reduce emissions. We consider several alternative cases, summarized in the three left-hand columns of Table 5.1 and annotated below:

Case (A): No elimination of emissions – base case.

Case (B): Similar to the 1982 legislation introduced by Senator Daniel Patrick Moynihan of New York – reduction in 31 states east of the Mississippi River.

Table 5.1. Predicted impact of emission reductions on rainfall acidity

Case	Reduction in emissions (%)		Expected acidity	
	x (%SO_2)	y (%NO_x)	$[H^+]$ (10^{-6} moles/liter)	pH
(A)	0	0	79.5	4.10
(B)	35	0	60.3	4.22
(C)	45	0	54.8	4.26
(D)	65	0	43.8	4.36
(E)	65	30	34.7	4.46
(F)	75	50	27.3	4.61
(G)	100	100	2.5	5.6

Case (C): Similar to the 1982 legislation introduced by Senator George Mitchell of Maine – reduction in 31 states east of the Mississippi River.

Case (D): Elimination of all SO_2 from utility boilers.

Case (E): Elimination of all SO_2 and NO_x from utility boilers.

Case (F): Elimination of all SO_2 and NO_x from utility and industrial boilers.

Case (G): Acidity of natural rain in equilibrium with atmospheric CO_2.

Each case, other than case (A), results in a reduction in the acidity of the acid rain. Alternatives (B) and (C) are illustrative of the many legislative proposals that were put forward in the early 1980s and are quite similar to the reductions eventually enacted in the 1990 Clean Air Act. Alternatives (D), (E), and (F) propose progressively deeper (and hence more costly) reductions in emissions. Finally case (G) is the "natural" background, requiring 100% elimination of acid emissions. The hydrogen ion concentrations that are estimated to result from each of the cases are given in the right-hand columns of Table 5.1 Note that the effect on pH is much less pronounced than the effect on the hydrogen ion concentration $[H^+]$ because the pH is calculated on a log scale.

Costs and Benefits of the Alternative Cases

Additional increments of emission reduction are progressively more costly to achieve; clearly it would cost a great deal more to eliminate all SO_2 and NO_x emissions from utility boilers (case (E)) than to eliminate SO_2 only (case (D)). A very rough estimate of the relative costs of the different cases is:

Case	% $[H^+]$ Reduction	Relative cost
(A)	0	0
(B)	25	10
(C)	31	15
(D)	45	50
(E)	56	100
(F)	66	200
(G)	97	∞

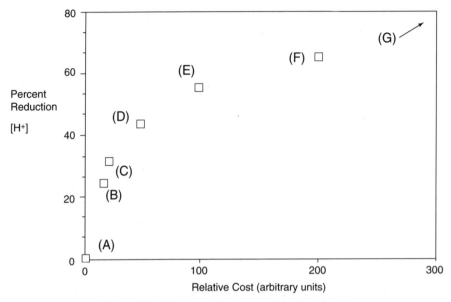

Figure 5.2. Emission reduction as a function of cost.

The resulting relationship between the desired benefit, which is the reduction in $[H^+]$ concentrations, and the cost is given in Figure 5.2. The relative costs used here are illustrative. (Indeed, the costs of cases (D), (E), and (F) are likely to be much greater than indicated.) Nevertheless, the curve clearly illustrates that the relationship between the desired benefit and the cost is not linear. Each increment of investment in pollution reduction brings diminishing returns in terms of a reduction in acid rain.

Data of the type represented in Figure 5.2 are very useful to decisionmakers. But they are not by themselves enough to determine the socially optimal amount of emission reduction. One reason is that there is no easy way to assign a monetary benefit to a given reduction in emissions, which would in turn allow a balancing of costs and benefits in monetary terms. Indeed, even obtaining a nonmonetary estimate of the environmental benefits is problematic. Although the sources of emissions are well known and the mechanisms of transport and deposition can be adequately modeled, the environmental impact of the acid rain deposition is much less well understood. In part, this is because there has not been extensive long-term monitoring that establishes a base line for the effects of acid rain. The impact of wet and dry acid deposition on soils and lakes is not known. The time required to reach a steady state in a lake or soil ecosystem is unknown, and the recovery potential of lakes, soils, and vegetation is also uncertain. In sum, whereas there is no doubt that reducing acid emissions will reduce acid deposition and thus reduce environmental damage, we cannot be confident about the precise degree of improvement. Nor can we be certain whether more subtle strategies – for example, reducing SO_2 emissions only when certain weather conditions prevail – would not be equally effective in reducing the environmental damage.

The Politics of Acid Rain

An even bigger obstacle to determining the optimal emission control strategy is the fact that the benefits and costs of pollution reduction are perceived and weighed differently by different groups. In such a situation, the question of "how much is enough" must be resolved through a political process.

High sulfur coal is mined in the states of Ohio, Illinois, Indiana, West Virginia, Kentucky, and Pennsylvania. These are also the states that have most of the coal burning utilities. Controls on acid rain would threaten the jobs of many miners and would also result in increased utility bills for consumers in these states. Not surprisingly, their political representatives fought strongly against many attempts to reduce emissions.

Environmentally minded politicians from the eastern states were the strongest advocates of more stringent regulation in Congress, and were often joined by supporters of other fuels, including low sulfur coals from the West. In addition to the conflicts between the East and the Midwest and between the western and midwestern coal producing states, there were many other hotly debated issues. Representative John Dingell of Michigan fought Representative Henry Waxman of California about provisions that regulated automobile emissions. There was also strong lobbying by Canada to control the acid rain that was affecting that country. And, as frequently happens, there was interagency competition between the Environmental Protection Agency, the Department of Energy, and NAPAP over which agency would have central responsibility for the many programs designed to improve understanding of acid rain or to develop new "clean coal technologies."

The initial clean air legislation enacted during the 1970s imposed uniform reductions on SO_2 emissions from all new coal power plants. The Clean Air Act of 1970 established a limit of 1.2 lb. of SO_2 emissions per million BTU of coal consumed. The New Source Performance Standards, adopted in 1977, required a uniform percentage reduction in SO_2 emissions from all new power plants, regardless of what kind of coal they were using. The effect was essentially to require the use of scrubbers on all new power plants, even those that could meet the 1970 standard of 1.2 lbs. of SO_2 per million BTUs. This was clearly not the most economically efficient solution. It did, however, serve the interests of the midwestern states, because it permitted the continued use of high-sulfur coal from those states, while making low-sulfur coal from the West relatively uneconomic in midwestern and eastern power plants since plant operators would have to pay both the extra costs of transportation and the cost of scrubbers.

A Market Mechanism for Pollution Control

The basic problem with setting uniform cleanup requirements on all plants is that some plants are less expensive to clean up than others, and from an economic efficiency point of view the sensible thing to do is to clean up those plants first, and then progressively move on to the higher-cost plants until the desired overall level of reduction is achieved.

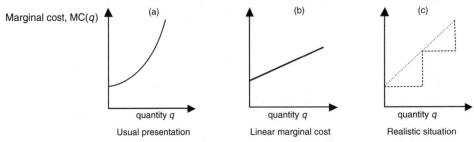

Figure 5.3. Marginal cost curves for emission reduction.

A helpful tool for analyzing the situation is the marginal cost curve for pollution control (see Figure 5.3). The marginal cost curve $MC(q)$ gives the cost for an additional unit of emission reduction at a given value of total emission reduction q. Ordinarily the marginal cost increases with q – each additional increment of pollution prevention is more difficult and hence costlier to achieve (another manifestation of the phenomenon of diminishing returns). In many situations the marginal cost increases exponentially with q (panel (a)), but for analytical convenience marginal cost curves are often assumed to be linear (panel (b)). (In practice marginal cost curves are typically discontinuous, as depicted in panel (c); each new pollution control measure involves significant investment, and provides a finite amount of emission reduction.)

The total cost required to reduce emissions by an amount q is

$$TC(q) = \int_o^q MC(q')dq'.$$

The average cost $AC(q) = TC(q)/q$ to reduce emissions by some amount q is evidently not equal to $MC(q)$. For a linear marginal cost curve we find

$$MC(q) = a + bq$$
$$TC(q) = aq + \frac{1}{2}bq^2$$
$$AC(q) = a + \frac{1}{2}bq.$$

Now suppose that government regulators want to reduce overall emissions from the industry by a total of Q units. To make things simple, suppose the industry consists of just two plants, A and B, with different marginal cost curves. The situation is illustrated in Figure 5.4. The graph is constructed so that the marginal cost curves for the two plants begin from opposite ends of the x-axis, and the length of the x-axis is Q units. This means that any point chosen along the x-axis divides the total of Q units of required emission reduction between the two plants. In this example, plant A has a steeper marginal cost curve than plant B, meaning that its cost of emission reduction is higher.

If the government wants Q units of reduction, the simplest scheme would be to require each plant to reduce its emissions by $Q/2$ units. This is shown in the left-hand panel of Figure 5.4. The total cost of pollution control in this case would be the sum of the shaded areas under the two curves. But a lower total cost strategy would be to

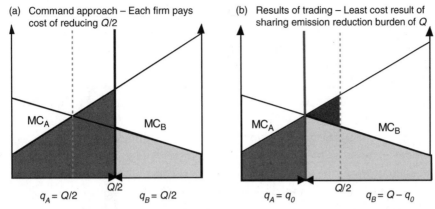

Figure 5.4. A simple two-plant illustration of the costs of regulatory compliance with emissions trading.

require plant A, whose costs of control are higher, to reduce its emissions by an amount $q_o(<Q/2)$, while plant B reduced its emissions by a larger amount $Q - q_o$. It is easy to show that the lowest total cost of emission reduction is achieved when q_o is given by the intersection of the two marginal cost curves (see panel (b)). The amount by which the total cost has been reduced relative to the equal reduction regulatory strategy is indicated by the cross-hatched area in panel (b). This result can be generalized to the case of an industry with many plants: In that case, the total cost of control is minimized when each plant's marginal cost is the same.

It is clearly to society's benefit to achieve the least cost result. But what mechanism is available to achieve this outcome? If the regulators knew the marginal cost curves of each plant, they could require each firm to reduce its emissions by an amount that would result in marginal costs being equalized across the industry. But this would require a depth of knowledge about each plant that government regulators do not generally have, and would entail a degree of regulatory micromanagement of individual plant operating decisions that would be fiercely resisted by the industry. An alternative approach is to create a market mechanism in which individual plants are assigned rights to emit a certain amount of the pollutant and then allowed to trade these rights with each other. Because of the tradability of these rights, plants where reduction would be most costly can purchase rights from other plants that can reduce their emissions less expensively. The result is the achievement of the desired goal at a lower cost than if all plants reduce emissions by the same amount, but in this case without the regulator needing to know the individual plants' marginal cost curves.

To see how this works, let us again use the simple two-plant model in Figure 5.4. To further simplify the model, let's suppose that each plant initially emits the same quantity, Z, of the pollutant in question (that is, before any regulations are applied.) As in the previous case, the regulator mandates an industry-wide reduction of Q units of emissions. But in this case the regulator issues each plant with $(Z - Q/2)$ emissions "allowances" – that is, each plant is given the right to emit $Z - Q/2$ units of the pollutant. These emissions allowances are tradable. Thus if plant A, whose marginal cost of control is higher, can purchase extra allowances from plant B at a lower price than the cost of reducing its own emissions, it will choose to do this. The additional allowances will enable it to emit more than its initial allocation of Z-$Q/2$ units. Conversely, plant B will be motivated to reduce its emissions by more than $Q/2$ if it can sell its unused allowances for a price higher than its marginal cost of emission reduction.

The trading of allowances between A and B will continue until the allowance price is just equal to the marginal cost of control for the two plants. At that point, B will have no incentive to make further reductions and sell the resulting unused allowances, and A will have no incentive to buy extra allowances instead of reducing its own emissions. Figure 5.4 shows that the total cost of achieving a reduction of Q units in this case is minimized, and is identical to that which would result if the regulators mandated the two plants to reduce their emissions by q_o and $Q - q_o$, respectively. But in this case the actual distribution of emissions between the two companies has been determined by a market mechanism rather than by regulatory fiat.

The allowance trading scheme can be generalized to the case of multiple plants. Each plant is given an initial allocation of emission allowances (which is smaller than the quantity of pollutant it would emit if unregulated). The plant considers the market price of allowances and then determines the least cost strategy for managing its emissions. This may involve some combination of implementing reductions and buying additional allowances. If, having reduced its emissions to a level equal to the number of allowances in its possession, the plant finds that the market price of allowances is above its marginal cost of control, it will undertake further reductions and sell its unused allowances into the market at a profit. The net result of all these individual strategies is a market clearing price for allowances that will ensure the least cost of accomplishing a given overall emission reduction target.

The Clean Air Act Amendments of 1990

In 1990 Congress adopted legislation that for the first time created a market mechanism for controlling emissions of SO_2 like the one described above. The Clean Air Act Amendments of 1990 established an aggregate ceiling (or "cap") on SO_2 emissions from all electricity plants, and introduced a system of tradable emission allowances designed to bring the utility industry below the cap at least cost.[2] Under the Act, after the year 2000 SO_2 emissions from the nation's power plants would be capped at about 9 million tons per year, down from the then-current level (in 1990) of about 23 million tons per year. (To set this in perspective, if 100 1000-MWe coal plants each burning 10,000 tons per day of 2% sulfur coal could be cleaned up to zero emissions, annual releases of SO_2 would be reduced by 14.6 million tons.)

The reduction was to be accomplished in two stages. The first phase targeted 263 of the largest, most-polluting coal-fired plants at 110 sites across the country and set an

---

[2] For NO_x emissions the new legislation continued to rely on "command and control" style emissions limits.

intermediate emissions cap to be implemented in 1995. In the second phase, the cap was reduced further and was extended to include all coal and oil-fired units. Every year each unit covered by the cap receives from the government a number of allowances according to a formula that takes account of the size of the unit and, for certain units, the age of the plant and the type of fuel and technology. (An allowance is defined as the right to emit one ton of SO_2 during one year.) The total number of allowances issued each year is equal to the aggregate annual emissions cap. Plant owners are then free to reduce their emissions below their allocation and sell their surplus allowances. Alternatively they may purchase additional allowances and emit a correspondingly larger amount of SO_2. Owners are also free to bank allowances for use in future years. At the end of the year each plant must have deposited enough allowances into an account maintained by the Environmental Protection Agency to cover all its recorded emissions in that year.

The Clean Air Act Amendments of 1990 were enacted only after a prolonged political struggle. Central issues included the questions of who should pay for the costs of emission reduction and how the interests of the miners in high-sulfur coal states could be protected. A special provision of the Act granted a two-year deadline extension to plants that elected to use scrubbers to reduce their emissions (as opposed to, say, switching to lower sulfur coals.) By (temporarily) favoring control technologies that allowed the use of high-sulfur coal, the provision provided some help to the mining industry in the Midwest and Appalachia. Critical to the passage of the Act was the support of important environmental organizations. Many in the environmental community were initially strongly opposed to the emissions trading mechanism, which they saw as giving power plants a 'license to pollute.' But in the ensuing debate some erstwhile opponents became convinced that the scheme could be effective in reducing emissions, and the eventual support of some important environmental groups was critical to the passage of the Act.

The SO_2 emissions trading scheme introduced by the Clean Air Act Amendments is generally considered to have been a success, achieving the aggregate emission reductions at lower cost than a command-and-control approach would have done. An MIT group has written an excellent book describing the program.[3] Predicting the total cost savings from emissions trading is a complex matter, but these authors credibly estimate that over a thirteen year period beginning in 1995 the market for SO_2 emission rights will reduce pollution abatement costs by $20 billion, to a level of approximately $15 billion. The utilities have clearly been taking advantage of the flexibility offered by the program. About 45% of the allowances issued in the first year of the program in 1995 were either traded between firms for use in that year or banked for later use.

The success of the SO_2 emissions trading program has stimulated interest in using similar schemes to control other pollutants, including greenhouse gases – the subject of the next chapter. To be effective, any such program requires a credible system of monitoring and enforcement. Emissions monitoring and enforcement are most easily

---

[3] A. Denny Ellerman, Paul L. Joskow, Richard Schmalensee, Juan-Pablo Montero, and Elizabeth Bailey, *Markets for Clean Air*, Cambridge University Press, Cambridge, United Kingdom, 2000.

carried out when there are relatively few emitting sources (as in the case of power plants), and when all those participating in the market for allowances are located within a single country. It should be noted that neither condition applies to the case of greenhouse gas emissions. Whatever the general applicability of the emissions trading scheme, however, the story of the Clean Air Act Amendments of 1990 is a good case study of how informed debate followed by government action can work to minimize the total cost to society of controlling a serious environmental externality.

6

Greenhouse Gases and Global Warming

Since the beginning of the industrial age, growing quantities of gases have been released into the atmosphere with the ability to trap sunlight and thus with the potential to cause an increase in the mean global temperature. A temperature increase of just a few degrees will lead to climate changes that have the potential to cause irreversible ecological impacts with enormous accompanying economic and social dislocations. The purpose of this chapter is to describe how the United States and other nations are dealing with this complex issue.

The quantity of gases for which human activity is responsible is small relative to both the total atmospheric inventory and the fluxes from natural sources such as plant growth and decay. As Figure 6.1 shows, the flux of carbon released today by the burning of fossil fuels is a very modest fraction of the carbon fluxes that are naturally exchanged between the atmosphere and the upper layers of the ocean and between the atmosphere and the terrestrial biosphere. But those natural flows had previously been in close balance, and the human contribution is growing rapidly (see Figure 6.2). This anthropogenic perturbation has the potential to destroy the delicate radiative balance that maintains the Earth's surface temperature.

Global warming is perhaps the most complex technology issue on the public policy agenda. The tasks of understanding the underlying science, predicting the climate impact of greenhouse gas emissions, and verifying these predictions all present extraordinary challenges. And because the adverse environmental effect here is global, the policy actions required to manage the problem must therefore be international in scope, vastly complicating the already difficult problems of balancing conflicting interests that arise in purely domestic environmental issues.

Unlike many of the other technology issues considered in this book, the global warming issue is still evolving. Our task in this chapter is to understand the nature of the scientific debate about global warming and the policy options for addressing the threat that it poses.

Basic Cause of the Greenhouse Gas Phenomenon

Solar radiation falls on the Earth with an energy distribution characteristic of a black-body radiating at the sun's surface temperature at roughly 6,000 degrees Kelvin. Most of this energy falls in the 0.2 to 4 micron range (1 micron $= 10^{-6}$ meters). The Earth absorbs this energy and re-emits it at longer wavelengths characteristic of a much cooler blackbody. Most of the re-emitted radiation falls in the 5 to 30 micron range.

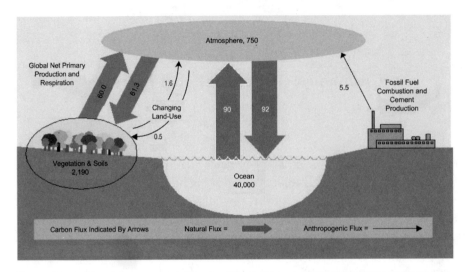

Source: Adapted from Intergovernmental Panel on Climate Change, at http:/www.grida.no/climate/vital/ 13.htm.

Figure 6.1. The global carbon cycle (all quantities are expressed in gigatonnes (10^9 metric tons) of carbon).

If the Earth had no atmosphere, this energy would be radiated directly into space, and a steady state energy balance between the two black body emitters predicts that the Earth's mean surface temperature would be 255 K ($-18°C$). But the presence of the atmosphere importantly affects this energy balance. The gases in the atmosphere are poor absorbers of short wavelength solar radiation, but some of them – principally water vapor – absorb some of the longer-wavelength radiation re-emitted by the Earth. The

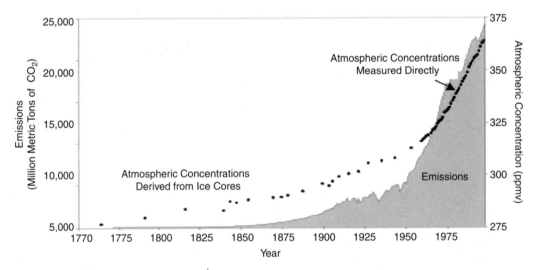

Source: Courtesy Carbon Dioxide Information Analysis Center, 2001.

Figure 6.2. Global emissions and atmospheric concentration of CO_2, 1750–1997.

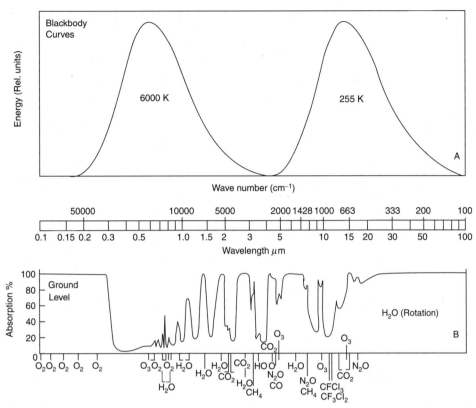

Source: Reprinted from J. F. B. Mitchell, "The 'Greenhouse' Effect and Climate Change," *Reviews of Geophysics*, 27 (1) (1989), 115–139.

Figure 6.3. Atmospheric absorption of radiation emissions from blackbodies. The upper panel shows the spectral distribution of emissions from blackbodies at 6,000K and 255K, corresponding to the mean emitting temperatures of the Sun and the Earth. The lower panel, showing the percentage of absorption for radiation passing through the atmosphere, indicates the region of weak absorption between eight and twelve microns in the thermal spectrum re-emitted by the Earth.

absorbed energy is re-radiated partly into space and partly back to the Earth's surface, thereby contributing to a net warming of the earth. The increase in the Earth's surface temperature resulting from the absorptive behavior of the atmosphere is what is known as the "greenhouse effect."

There is, however, a "window" in the infrared region, roughly between 8 microns and 12 microns, in which re-absorption of energy by the gases that are naturally present in the atmosphere is very weak. Much of the re-emitted radiation in this range thus passes directly through the atmosphere and escapes the Earth (see Figure 6.3).

If chemical species that absorb significantly in the 8 to 12 micron window are introduced into the atmosphere, the radiative balance of the Earth could be affected – more energy would be absorbed and the mean global temperature might then rise. Important gases with this property include carbon dioxide (CO_2), chlorofluorocarbons (CFCs), methane (CH_4), nitrous oxide (N_2O), and tropospheric ozone. These are the

Table 6.1. Important greenhouse gases

Species	Atmospheric concentration (parts per million by volume)	Global warming potential[†]
Nitrous oxide (N$_2$O)	0.28	296
Methane (CH$_4$)	1.7	23
CFC-12 (CCl$_2$F$_2$)	5×10^{-4}	6,200–7,100*
HCFC-22 (CHClF$_2$)	1×10^{-4}	1,300–1,400*
Perfluoromethane (CF$_4$)	7×10^{-5}	5,700
Sulfur hexafluoride (SF$_6$)	3×10^{-5}	22,200
Carbon dioxide (CO$_2$)	358	1

[†] Calculated for 100-year time horizon
* Includes indirect effects due to ozone depletion.

Source: Intergovernmental Panel on Climate Change, *Climate Change 2001: The Scientific Basis*, Cambridge University Press, Cambridge, UK, 1996, p. 38 and 338–9, reproduced in U.S. Energy Information Administration, *Emissions of Greenhouse Gases in the United States 2000*, at http://www.eia.doe.gov/oiaf/1605/ggrpt/tbl3.html (visited August 28, 2002).

so-called "greenhouse gases" (a list that strictly speaking should also include water vapor, because of its dominant contribution to the greenhouse effect in other parts of the spectrum.) The quantitative effect of such chemical species will, of course, depend on a number of factors, including: (1) the atmospheric concentrations of the species; and (2) their absorptive strength. For example, one molecule of freon, CF$_2$Cl$_2$, is about 10,000 times more effective at trapping radiation in the window than a molecule of CO$_2$.[1]

The current concentrations of greenhouse gases in the upper atmosphere are given in Table 6.1 Also shown in Table 6.1 is the "Global Warming Potential" of each gas, which is an index of the warming effect of an instantaneous emission of 1 kg of the gas relative to 1 kilogram of CO$_2$, where the effects are analyzed over a specified time horizon (in this case 100 years).

Table 6.2 gives the Department of Energy's estimate of U.S. emissions of greenhouse gases from anthropogenic sources in 1998. The estimates are expressed in millions of metric tons of carbon equivalents. To obtain these numbers, the emission of each gas (in metric tons) is weighted by its Global Warming Potential, with CO$_2$ having a weight of 1. The CO$_2$ equivalents are then converted to carbon equivalents by multiplying by $\frac{12}{12+2\times16} = 0.273$. It is noteworthy that the United States is responsible for about one quarter of the world's greenhouse gas emissions.[2]

---

[1] A rough indication of the relative radiative contribution of a chemical species in the atmosphere is obtained as follows: Let $c(t)$ denote the concentration of a compound in the atmosphere that is added at a rate q. The compound has an average lifetime in the atmosphere of τ; in that period either precipitation or chemical reaction removes the compound. Accordingly,

$$\frac{dc(\tau)}{dt} = -\tau^{-1}c(t) + q$$

and in steady state $c_s = q^\tau$. If the absorption coefficient is α then the relative radiative contribution from this species is $\alpha\tau q$.

[2] In 2000, the U.S. emissions of carbon dioxide from the consumption and flaring of fossil fuels totaled 1,571 million metric tons of carbon equivalent, out of a world total of 6,443 million metric tons (U.S.

Table 6.2. Estimated U.S. emissions of greenhouse
gases – 1998

	Million metric tons of carbon equivalent
Carbon dioxide	1507
Methane	168
Nitrous oxide	103
HFCs, PFCs, SF_6	40

Source: Energy Information Administration, *Annual Energy Review: Environmental Indicators*, http://www.eia.doe.gov/emeu/aer/envir. html (updated May 7, 2002, visited August 28, 2002).

Calculating the Effect of Greenhouse Gases on the Climate

When the atmospheric concentration of a greenhouse gas is increased, a variety of direct and indirect responses will influence the net amount of energy that is retained. First, it is necessary to consider other species that absorb in the same wavelength band. If the concentrations of these are high enough, absorption in the relevant range may be saturated. Then, if the species are chemically reactive, it is also necessary to consider the effects of atmospheric chemistry on the concentrations. And third, an initial increase in energy absorption may trigger a range of feedback mechanisms, both positive (+) and negative (−), such as:

- an increase in blackbody cooling to space   (−)
- an increase in atmospheric water vapor   (+)
- a reduction in snow and ice coverage   (+)
- an increase of cloud cover   (+)
- an increase in the water content of clouds   (−)

Calculating the temperature response to an increase in the concentration of any one greenhouse gas is thus a very complicated problem.

Global Climate Models

In order to make a quantitative estimate of the global response to a change in the emission of a greenhouse gas, it is necessary to have a model that describes the complicated chemistry and radiative physics of the atmosphere with feedback. This type of model is called a general circulation model (GCM). Three-dimensional general circulation models solve for the local wind velocity, temperature, pressure, and humidity at each point on a three-dimensional grid that is superimposed on the atmospheric continuum. The GCM models generate numerical solutions to the nonlinear Navier–Stokes equations, augmented by appropriate equations to describe solar heating, important chemical reactions, and radiation and thermal transport. The system is inhomogeneous because

Energy Information Administration, at http://eia.doe.gov/emeu/iea/tableh1.html (updated on April 24, 2002, visited August 28, 2002).

the presence of ice, clouds, snow, and aerosols must be considered. Boundary conditions are complex: the radiation absorbed and emitted by the ocean surface (partially covered by sea ice) and land surface (covered by vegetation, ice, or snow) vary by season. Atmospheric circulation is also critically affected by the circulation of the oceans, which is in turn driven by the transport of heat and moisture and momentum from the atmosphere. The absorption of CO_2 by the oceans is another extremely important aspect of the interaction between the atmosphere and the oceans. The computational demands made by coupled three-dimensional ocean and atmospheric circulation models are extraordinary.

Today the best that can be done is to use a model grid with a dimension of a few hundred miles – far larger than the characteristic scale of key phenomena involved in atmospheric processes such as clouds, sea ice, vegetation, and snow. Yet even this overly coarse grid imposes numerical demands that only a handful of the world's largest and most powerful computers are capable of meeting. (Laboratories able to carry out such simulations include the NASA Goddard Laboratory, Princeton University's Global Fluid Dynamics Laboratory, the Department of Commerce's National Center for Atmospheric Research, the Department of Energy's Lawrence Livermore National Laboratory, and the United Kingdom Meteorological Office).

There is an additional technical complication in the modeling of global climate change. The nonlinear phenomena that a GCM attempts to describe exhibit chaotic behavior, that is, fluctuating values of dynamical variables that appear to be random but that in fact accurately follow precise nonlinear dynamics of the system. The trajectories from slightly different initial conditions are enormously different. Moreover, the system can make abrupt transitions from one stationary state to another. The turbulent character of wind and variable weather patterns are reflections of this chaotic behavior.

There are two important implications of the nonlinear character of climate systems. First, the finite spatial and temporal mesh employed in the computer simulations can lead to very different realizations of some important phenomena, for example, the spatial distribution of cloud formations. Second, the fluctuating character of both model predictions and empirical observations means that it is difficult to calibrate and verify the GCM predictions. The natural variability of the climate means that the detection of anthropogenic climate changes is a particularly difficult problem.

This difficulty in verifying model predictions is not unique. There are many other situations in public policy in which model predictions are used as a basis for policy when there is not the slightest shred of empirical evidence to verify the model's validity. An excellent example is the use of large scale simulations of nuclear weapons exchanges during the Cold War to determine the size of the nuclear deterrent force of the United States. There was obviously no empirical verification of the models in that case, but people agreed anyway that the model results were credible. In the case of global warming, however, the difficulty of verifying the models, combined with the fact that the anthropogenic sources of greenhouse gases are only a small fraction of the total flux

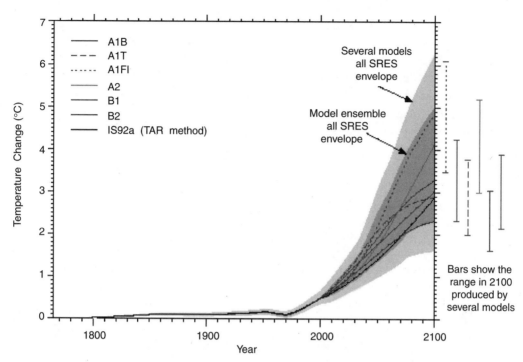

Source: Reprinted from Intergovernmental Panel on Climate Change, *Technical Summary of the Working Group I Report*, p. 61, at http://arch.rivm.nl/env/int/ipcc/images/wgl_ts.pdf (updated on June 3, 2002, visited on August 29, 2002).

Figure 6.4. IPCC projections of global mean temperature increases for illustrative greenhouse gas emission scenarios (SRES scenarios).

exchanged with the atmosphere, are obstacles to the achievement of broad consensus on the magnitude and timing of the environmental effects.

Predictions of Global Climate Models

The closest approximation to a consensus prediction of the global warming models today is provided by the Intergovernmental Panel on Climate Change (IPCC), an international group of climate scientists established by the United Nations and the World Meteorological Organization in 1988. Working Group I of the IPCC, charged with assessing the science of climate change, issued its third major assessment of the field in 2001. The Third Assessment Report found that:

> By 2100, carbon cycle models project atmospheric CO_2 concentrations of 540 ppm to 970 ppm for the illustrative emissions scenarios (90% to 250% above the concentration of 280 ppm in 1750). . . . These projections include the land and ocean climate feedbacks.[3]

Figure 6.4, taken from the same report, illustrates the range of global mean temperature increases projected to occur between 1990 and 2100. The six curves represent

[3] Intergovernmental Panel on Climate Change, *Technical Summary of the Working Group I Report*, p. 63, at http://arch.rivm.nl/env/int/ipcc/images/wgl_ts.pdf, updated on June 3, 2002.

six illustrative greenhouse gas emission scenarios. (A seventh curve – IS92a – is an older scenario included for comparison purposes.) The shaded areas represent the envelope of the full set of thirty-five emissions scenarios considered by the IPCC. (The lighter shaded area is the envelope of projections from seven different climate models; the darker shading uses the average of the results from the seven models.) In sum, a rate of temperature increase of approximately 0.2°C to 0.5°C per decade is expected over the next century. By comparison, the global average surface temperature increased by about 0.6°C since the late 19th century. The IPCC report concluded that

> The globally averaged surface temperature is projected to increase by 1.4°C to 5.8°C over the period 1990 to 2100. These results are for the full range of 35 emission scenarios [considered by the IPCC], based on a number of climate models. . . . The projected rate of warming is much larger than the observed changes during the 20th century and is very likely to be without precedent during at least the last 10,000 years, based on palaeoclimate data. . . . [4]
>
> It is very likely that nearly all land areas will warm more rapidly than the global average, particularly those at northern high latitudes in the cold season . . . in winter the warming for all high-latitude northern regions exceeds the global mean warming in each model by more than 40% (1.3°C to 6.3°C for the range of models and scenarios considered.)[5]

Although the IPCC assessment has been sharply criticized by some (for example, the George C. Marshall Institute), it stands today as the best statement of consensus among the knowledgeable scientific community. The extent of the consensus has led several nations, especially Canada and smaller European countries, to call for policy measures to cap and reduce the present level of emissions.

From the viewpoint of social, ecological, and economic impact, regional variations in the temperature increase are more important than the global mean temperature increase, since certain regions of the world are much more sensitive to slight changes in temperature and humidity than others. GCM predictions indicate significant geographical and seasonal variation. But in general they predict greater rainfall, drier soil, an increase in the sea level, and a reduction in sea ice and polar ice caps.

The net effect of such changes on agriculture, natural ecosystems, and human life patterns is not addressed by the climate models. The severity of the impact will depend on the length of time and the path that the climate system takes to reach a new steady state. If the changes occur gradually over centuries it is more reasonable to suppose that human and natural systems will adjust. If significant changes occur abruptly, there will be no opportunity for a gradual transition and the impacts could be severe.

[4] Intergovernmental Panel, op. cit., p. 69.
[5] Intergovernmental Panel, op. cit., p. 69.

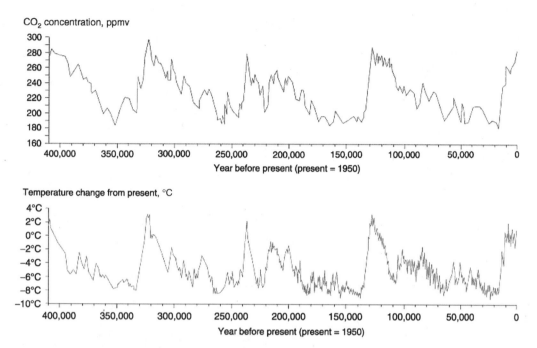

Source: Reprinted from J. R. Petit et al, "Climate and atmospheric history of the past 420,000 years from the Vostok ice core in Antarctica," *Nature* 399, (1999), 420–436.

Figure 6.5. Temperature and CO_2 concentration in the atmosphere over the past 400,000 years (from the Vostok ice core).

What is Known Empirically

Air bubbles trapped in ancient polar ice can be analyzed to determine the composition of the atmosphere hundreds of thousands of years ago. Ice core samples taken from Vostok in Antarctica reveal a striking correlation between temperature change and carbon dioxide composition over hundreds of thousands of years (see Figure 6.5). The correlation between temperature and carbon dioxide can also be seen in the record of the last century as Figure 6.6 shows. During this period, the atmospheric concentration of carbon dioxide rose steadily by about 30%, while the mean global temperature rose by about 0.6°C. However, there is high variability of surface temperature over five-year increments, and the correlation is much weaker on this time scale. The decade of the 1990s seems likely to have been the warmest since the instrumental record began in 1861.

The key question is the role of human activity in the recent hundred-year warming trend. Alternative explanations such as natural temperature variations, solar radiation fluctuations, and volcanic emissions must also be considered. Although the evidence remains somewhat equivocal that the recent warming trend bears the fingerprint of human influence, doubts on this score are dissipating. In its latest report, the IPCC

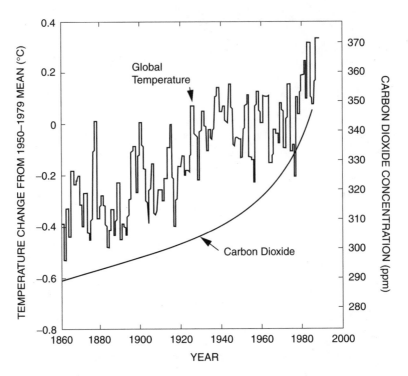

Source: Reprinted from S. H. Schneider, "Prediction of Future Climate Change," in *Energy and the Environment*, ed. J. Tester et al. (Cambridge, MA: MIT Press, 1991), 50.

Figure 6.6. Temperature and CO_2 concentration in the atmosphere, 1860–2000.

reached the following conclusion:

> There is new and stronger evidence that most of the warming observed over the last 50 years is attributable to human activities.[6]

Conclusions Concerning Global Climate Modeling

The issue of greenhouse gases and climate change cannot be addressed without the use of global climate models. The model predictions are extremely sensitive to input assumptions, and verification is not a simple matter. There is general agreement that general circulation models are relatively successful at predicting seasonal and large-scale spatial and temporal changes. However, critics point out that the models cannot "run backwards." From presently known conditions, the models cannot describe the recent past. Today, moreover, GCM cannot quantitatively describe climate variations on a regional basis over a time period of less than several decades. Finally, certain important phenomena are widely agreed to be poorly described by the models, for

[6] Intergovernmental Panel on Climate Change, *Technical Summary of the Working Group I Report*, p. 61, at http://arch.rivm.nl/env/int/ipcc/images/wgl_ts.pdf (updated on June 3, 2002, visited on August 29, 2002).

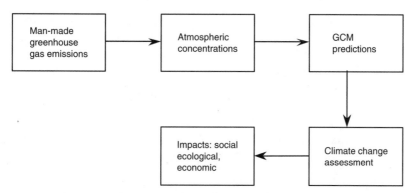

Figure 6.7. The scientific chain of reasoning for global warming.

example, cloud formation, aerosol behavior, and the absorptive capacity of the oceans for CO_2.

Of course, the GCM predictions are only one step in the sequence required to assess the consequences of global climate change. The overall chain of reasoning is depicted in Figure 6.7. Each step in this sequence requires extensive analysis and is subject to significant uncertainty from the perspective of making policy.

POLICY ISSUES POSED BY GLOBAL WARMING

The fundamental policy issue posed by global warming is: "What do we know, and what should we do about it?" These are questions of great significance. The harm from changes in the earth's climate is potentially enormous and the consequences of doing nothing may be very great. On the other hand, the cost of pre-emptive measures will surely be very high. The analogy to insurance is tempting. Shouldn't we buy some protection today to avoid the consequences of a possible catastrophe in the future? This qualitative notion is reasonable enough, but it does not point to the amount of insurance to buy or to the mechanism for providing protection against the worst, albeit improbable, consequences.

How Costly are Mitigating Strategies?

The Japanese engineering professor Yoichi Kaya and his colleagues adapted an earlier formulation by John Holdren and Paul Ehrlich to illustrate how difficult it will be to cap or reduce CO_2 emissions.[7] Consider the following identity that relates CO_2 emitted by fossil fuel use (C), energy produced (E), economic output (Y), and population (P). The identity is:

$$C = (C/E) \times (E/Y) \times (Y/P) \times (P).$$

This identity can be applied to any region or to the entire world. Because it is an

---

[7] Y. Kaya, K. Yamaji, and R. Matsuhashi, "A grand strategy for global warming," paper presented at the Tokyo Conference on the Global Environment, September 11–13, 1989.

accounting identity, it must be true. It stipulates that the goal of reducing the quantity of CO_2 emissions, that is the value of C on the left-hand side of the equation, imposes a constraint on the quantities on the right-hand side of the equation.[8]

The nature of the constraint is revealed by differentiation of the equation:

$$\frac{\delta C}{C} = \frac{\delta(C/E)}{(C/E)} + \frac{\delta(E/Y)}{(E/Y)} + \frac{\delta(Y/P)}{(Y/P)} + \frac{\delta P}{P},$$

where the differential value δX refers to a change in the quantity or ratio X. Thus the constraint is that within any given geographic area, for any specified time interval, the sum of the fractional changes of the quantities on the right hand side of the equation must equal the fractional change in the carbon emitted.[9]

An alternative formulation of the identity would not include the population dependence

$$C = (C/E) \times (E/Y) \times Y$$

that in turn leads to the relation

$$\frac{\delta C}{C} = \frac{\delta(C/E)}{(C/E)} + \frac{\delta(E/Y)}{(E/Y)} + \frac{\delta Y}{Y}$$

revealing a simpler (but still uncomfortable) constraint.

The accounting identity provides no information as to whether any particular scenario is actually possible. But it reveals what is impossible and shows the constraints on the rate of change of the chosen variables. (The identity holds even if there are random events in a certain time period, e.g., volcanic eruptions.) Here are some scenarios that illustrate how informative the Kaya identity can be:

1. *"Traditional growth" scenario.* Here we assume no reduction in the carbon intensity of the energy system and no improvement in energy efficiency. Thus,

$$\delta(C/E) = \delta(E/Y) = 0.$$

If there is, in addition, vigorous economic growth and continued population growth, for example,

$$\frac{\delta(Y/P)}{Y/P} = 3\%/\text{yr} \qquad \frac{\delta P}{P} = 2\%/\text{yr},$$

---

[8] The identity does not account for nonenergy-related sources of CO_2. But because about 80% of all anthropogenic emissions of carbon dioxide currently come from fossil fuel combustion the focus on energy use is appropriate. The identity also doesn't consider other greenhouse gases. The significance of this omission is discussed later in the chapter.

[9] Note that the differential form is quite accurate:
Consider $Y = (Y/C) \times C$, then:
$$Y_2 = (Y/C)_2 C_2$$
$$\frac{\Delta Y}{Y_1} = \frac{(Y/C)_2 C_2 - (Y/C)_1 C_1}{Y_1},$$
so
$$\frac{\Delta Y}{Y_1} = \frac{\Delta(Y/C)}{(Y/C)_1} + \frac{\Delta C}{C_1} + \frac{\Delta(Y/C)}{(Y/C)_1}\frac{\Delta C}{C_1}.$$

the consequence will be considerable growth in carbon emissions:

$$\frac{\delta C}{C} = 5\%/\text{yr}.$$

2. *"Developed country" scenario.* We assume in this case that there will be no population growth, and also impose a requirement that carbon emissions do not grow. Therefore,

$$\delta C = \delta P = 0.$$

If, in addition, these countries achieve annual energy efficiency gains of 1% and annual reductions in carbon intensity of 1%,

$$\frac{\delta(C/E)}{(C/E)} = \frac{\delta(E/Y)}{(E/Y)} = -1\%/\text{yr},$$

then per capita income growth must be:

$$\frac{\delta(Y/P)}{(Y/P)} = 2\%/\text{yr}.$$

3. *"Developing country" scenario.* Developing countries such as India are experiencing population growth, an increase in energy intensity because of the changing composition of their industrial activity, and no reduction in carbon intensity. We assume

$$\frac{\delta P}{P} = 2\%/\text{yr} \quad \frac{\delta(C/E)}{(C/E)} = 0 \quad \frac{\delta(E/Y)}{(E/Y)} = 1\%/\text{yr},$$

so they face the constraint

$$\frac{\delta C}{C} = 3 + \frac{\delta(Y/P)}{(Y/P)} \ \%/\text{yr}.$$

4. *Estimates for the United States by the Department of Energy.* Table 6.3 presents estimates of what the United States could do to reduce its carbon emissions over the next two decades under various economic growth and energy supply and demand assumptions. The estimates were prepared for the DOE in 1997 by the Interlaboratory Working Group on Energy-Efficient and Low-Carbon Technologies (the so-called "Five Lab Study"). The study created four scenarios of progressively increased energy efficiency and lower carbon emissions using information about the availability, performance and costs of energy efficiency and supply technologies. All four scenarios assumed the same overall economic growth rate. The estimates of total carbon emissions for each scenario were built up from sector-level models of technology substitution in the building, industrial, transportation, and utility sectors. These "bottom-up" estimates can be compared with the results obtained by applying the Kaya identity to the assumptions of the study – that is, by requiring macro-level self-consistency in each of the scenarios. As Table 6.3 shows, the agreement is quite good, although not perfect.

The Kaya accounting identity sheds useful light on the obligations undertaken by most developed countries under the 1997 Kyoto Protocol of the U.N. Framework Convention on Climate Change. Many of these countries committed to reduce their GHG

Table 6.3. Average annual energy and carbon emission growth rates for the United States, 1997–2010

	Business-as-usual	Efficiency case	High efficiency/low carbon case ($25/tonne)	High efficiency/low carbon case ($50/tonne)
Gross domestic product (GDP)	1.88%	1.88%	1.88%	1.88%
Energy demand	1.09%	0.56%	0.34%	0.13%
Energy consumption per GDP	−0.77%	−1.30%	−1.51%	−1.71%
Carbon emissions per GDP	−0.63%	−1.20%	−1.73%	−2.58%
Carbon emissions	1.24%	0.65%	0.11%	−0.75%
Carbon emissions required by Kaya identity	*1.22%*	*0.67%*	*0.13%*	*−0.73%*

Source: Interlaboratory Working Group on Energy-Efficient and Low-Carbon Technologies, *Scenarios of U.S. Carbon Reductions*, prepared for the Office of Energy Efficiency and Renewable Energy, U.S. Department of Energy, 1997.

emissions by 6% to 8% below the 1990 level by roughly 2010 – for the United States the target was 7%. (The requirements of the Kyoto Protocol are discussed in more detail later in the chapter.) The difficulty of achieving this target will depend partly on the magnitude of the increase in GHG emission rates that occurred between 1990 and 1997. For the United States, GHG emissions rose by about 13% during this period, so the commitment to a 7% cut implies reductions averaging approximately 1.8% per year between 1997 and 2010. If we further assume that the United States will achieve a 1.5% per year improvement in energy efficiency and a 1% per year reduction in carbon intensity during this period, the Kyoto target constrains growth in gross domestic product to less than 1% per year. Most countries – the United States included – would not find an economic growth rate of 1% per year acceptable. (Economic growth in the United States during the 1990s actually exceeded 3% per year.) Kaya's identity imposes an intellectual discipline that is often lacking in policy discussions of the greenhouse gas issue.

What are the Policy Choices?

The policy response to global climate change hinges on a judgment about the magnitude, timing, and cost of climate changes compared to the cost of reducing greenhouse gas emissions from energy use. Broadly speaking, there are three policy positions.

The first position is to do nothing. It is intellectually respectable to believe that too little is currently known about global climate change from greenhouse gases to take costly policy measures. In this view, the uncertainties about the timing of a temperature increase, the magnitude of the increase, and most importantly the economic and environmental consequences of the resulting climate change are sufficiently large to support a position of "no action now." This position does not necessarily imply a rejection of the consensus scientific view that increasing GHG concentrations will lead to global warming and that anthropogenic GHG emissions are largely responsible for the increasing atmospheric concentration of GHGs.

The strategist Tom Schelling offers the following argument for doing nothing.[10] He notes that the adverse effects of global climate change will occur many years in the future, perhaps 75 or 100 years from now, and that the effects may not be nearly so harmful as currently expected. He further argues that present environmental dangers (such as toxic wastes in Eastern Europe) are more deserving of resources today, and that these resources should not be diverted to reducing the environmental burdens on future generations. Future generations, if they had a voice today, might well prefer that we invest resources now in economic growth so that more resources will be available in the future to meet the needs that will exist at that time.

The second position is to do the easy things. Easy actions are those that almost everyone supports, that do not cost much money, and that may be justifiable on other grounds besides. In the case of global warming the easy options include:

- Banning emissions of the most damaging greenhouse gases, the chlorofluorocarbons e.g., $CFCl_3$ and CF_2Cl_2. These gases are especially strong absorbers of long wavelength infra-red radiation and they are also implicated in destruction of the stratospheric ozone layer. The Montreal Protocol on Substances that Deplete the Ozone Layer, which entered into force in 1989, did in fact take this action.
- Continuing a vigorous research program on the causes and consequences of global warming and ensuring that this knowledge is available to other nations. In recent years, the U.S. government has been willing to expand considerably the research effort on global warming. The effort is somewhat hampered, however, by the absence of a single lead agency with responsibility for global warming. There is bureaucratic competition for the available research support, with each agency pursuing an approach that is consistent with its mission. Thus, for example, NASA is developing an elaborate space-based "earth observation system" that will permit atmospheric measurements from space; NOAA, the National Oceanic and Atmospheric Administration, is sponsoring the development of general circulation climate models, and so on.

Other relatively easy options involve policy measures that seek to increase energy efficiency or reduce carbon intensity. There are several possibilities. Most experts agree that the most important step is to ensure that energy is realistically priced; so governments should eliminate subsidies designed to keep energy prices low, and should let prices reflect the workings of the marketplace. Subsidized energy prices effectively increase energy use and hence environmental emissions, including greenhouse gases. Energy taxes reduce the demand for fuels and thus, if applied selectively, can be used as

[10] Thomas Schelling, comments at Plenary Session II, in J. Tester et al (eds.), *Energy and the Environment in the 21th Century*, MIT Press, Cambridge, MA, 1991, p. 148–151.

Table 6.4. Low or zero cost greenhouse gas mitigation options for
the United States

Sector	Estimated reduction in GHG* (millions of tons of CO_2 equivalent)
Residential and commercial energy management (lighting, heating, cooking, refrigeration, cooling, ventilation, etc.)	890
Industrial energy management	527
Transportation energy management	290
Electricity and fuel supply	47
TOTAL	1,754

* The emission reduction estimates assume 100% penetration of the market.

Source: National Academy of Sciences, *Policy Implications of Greenhouse Warming: Mitigation, Adaptation, and the Science Base*, National Academy Press, Washington, D.C., 1992, p. 477–8.

an incentive to switch fuels from coal and oil to natural gas or to noncarbon fuels like nuclear and renewable energy technologies. European countries impose much higher taxes on energy, especially gasoline, with resulting improvements in transportation energy efficiency and greater incentives for new automobile technologies.

Research and development directed towards increases in energy efficiency and reductions in carbon intensity is of interest to the United States and other nations for many reasons beyond global warming. Greater energy efficiency means fewer environmental effects of other kinds as well as less use of petroleum – a depleting resource upon which the United States and other nations depend too heavily, especially imports. The problem is to identify those actions that are most cost-effective in reducing carbon efficiency and to determine the policy measures and incentives required to achieve the desired results.

Several recent studies have attempted to identify the most cost-effective ways of decreasing GHG emissions. A 1991 study by the National Academy of Sciences identified significant opportunities for reducing GHG emissions in the United States simply by undertaking cost-effective energy conservation.[11] The Academy panel concluded that large reductions could be realized by introducing improved energy efficiency and energy management techniques costing little or nothing or even generating net savings (see Table 6.4).

The panel also considered additional energy efficiency measures and estimated the cost of each increment using a 6% real discount rate. A simplified representation of this result is given in Figure 6.8. The importance of the figure is that it relates incremental improvements in energy efficiency to incremental costs. It gives an indication of how much benefit, in terms of emission reductions, could in principle be purchased at a given price. This is an example of a "bottom-up" analysis of the economic impacts of climate protection policies. Bottom-up analyses consider the technological options that are available for reducing emissions in different parts of the economy and then

[11] National Academy of Sciences, *Policy Implications of Greenhouse Warming: Mitigation, Adaptation, and the Science Base*, National Academy Press, Washington, D.C., 1992.

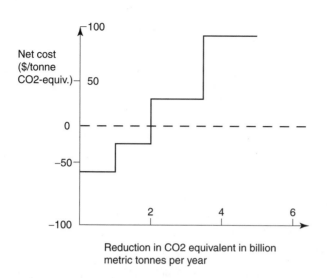

Source: Reprinted from National Academy of Sciences, *Policy Implications of Greenhouse Warming: Mitigation, Adaptation, and the Science Base* (Washington, DC: National Academy Press, 1992).

Figure 6.8. The relationship between the cost of energy efficiency measures and the potentially achievable reductions in U.S. carbon dioxide emissions. (A broad band should be placed around the step curve to indicate uncertainty as to the market penetration of the energy conservation measure.)

estimate the overall cost of achieving a given level of reduction by adding up the costs and contributions of the different measures. In contrast, "top-down" analyses of economic impacts are based on models of the entire economy, which simulate aggregate consumption, savings, and investment, together with the tax, spending, and monetary policies of the government. We will have more to say about the calculation of economic impacts later in the chapter.

The third policy position is to adopt tough measures. The proposal most frequently put forward is to levy a carbon tax. A tax on carbon emissions (as well as on other greenhouse gases) would create an incentive to find alternatives that emit less carbon. William Nordhaus of Yale University estimates that a $5 per tonne carbon tax would reduce CO_2 emissions by 13% while raising coal prices by about 10% and oil prices by about 3%.[12] The economic cost of the carbon tax is the reduced output and lower growth that would accompany higher energy prices. However, the manner in which the tax revenue is used can compensate to some extent for these adverse economic effects. The level of tax proposed by Nordhaus is not terribly burdensome. He argues that this level of taxation is sufficient because of his relatively low estimates of the adverse economic impacts of future global climate change. However, he also notes that a 50% reduction in the level of CO_2 emissions would require a tax of about $130 per tonne of CO_2 and that the net resource cost of this tax would be about $180 billion annually, or about 1% of world output – by no means a trivial economic burden. (Note: A tax of $100 per tonne of carbon in the United States would generate revenues of about $140 billion per year. Such a tax would increase the price of gasoline by about $.25 per gallon.)

[12] William D. Nordhaus, "Economic policy in the face of global warming," in *Energy and the Environment in the 21st Century* op. cit. p. 103–118.

Table 6.5. World carbon dioxide emissions from the consumption and
flaring of fossil fuels, 2000

	Million metric tons of carbon equivalent	Percentage share
North America	1,832	28.4
Central and South America	269	4.2
Western Europe	1,000	15.5
Eastern Europe and Former USSR	844	13.1
Middle East	288	4.5
Africa	240	3.7
Asia & Oceania	1,970	30.1

Source: Energy Information Administration, International Energy Outlook, Table H1,
http://www.eia.doe.gov/emeu/iea/tableh1.html (updated on April 24, 2002, visited August
30, 2002).

Other candidates for the "tough measures" category include climate engineering
schemes for actively offsetting the climate effects of greenhouse gas emissions. These
include (1) reforestation and pickling trees; (2) painting roads and roofs white (to reflect
sunlight back into the atmosphere before it is absorbed on Earth); (3) adding particles
to the stratosphere to reflect the incoming solar radiation; and (4) building dikes to
keep out higher ocean tides. Some of these climate engineering options raise significant
environmental issues of their own.

International Implications

Global warming, by definition, is a global problem. Little good will flow from GHG
emission reductions in some countries if others are increasing theirs. Benefits will only
be realized if total emissions are reduced. So management of the problem entails reaching
international consensus on what needs to be done and then deciding on a procedure
for allocating the economic burdens of reductions among different nations.

Table 6.6. The world's largest carbon
dioxide emitters, 2000*

	Million metric tons of carbon equivalent	
United States	1,571	(24.3)
China	775	(12.0)
Russia	450	(7.0)
Japan	314	(4.9)
India	253	(3.9)
Germany	229	(3.6)

*From the flaring and consumption of fossil fuels.
Source: Energy Information Administration, *International Energy Outlook*, Table H1, http://www.eia.doe.
gov/emeu/iea/tableh1.html (updated on April 24, 2002,
visited August 30, 2002).

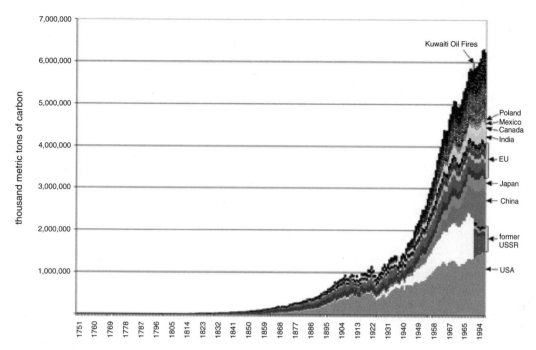

Source: Reprinted from G. Marland, T. A. Boden, and R. J. Andres at http://cdiac.esd.ornl.gov/ftp/ndp030/nation98.ems (updated July 25, 2001, visited August 30, 2002).

Figure 6.9. Global CO_2 emissions from fossil-fuel burning, cement manufacture, and gas flaring, 1751–1996.

The current regional distribution of carbon emissions from the use of fossil fuels is shown in Table 6.5, and the largest national emitters are listed in Table 6.6. The United States alone accounts for 24% of the total, and the world's other advanced economies account for another 24%. These developed economies have been responsible for most of the carbon dioxide emissions since the beginning of the industrial age, as Figure 6.9 shows.

In the future, however, most of the growth in GHG emissions is expected to come from the developing world, where much of the world's population is located and where the aspirations for economic growth are high. The U.S. Energy Information Administration projects that energy consumption by the developing nations – notably including China, India, Indonesia, and Brazil – will overtake that of the advanced economies by about 2020 (see Figure 6.10). Clearly the developing world and developed world have different interests, and there is plenty of room for controversy.

THE INTERNATIONAL DIPLOMATIC PROCESS ON CLIMATE CHANGE

Canada and the smaller European nations have been calling for a more vigorous international approach to the problem of controlling greenhouse gas emissions. Other developed countries, especially the United States, have been reluctant to take costly

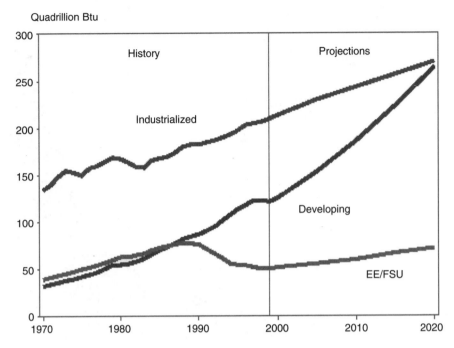

Source: Reprinted from Energy Information Administration, *International Energy Outlook – 2002*, Table 15, 9, at http://www.eia.doe.gov/oiaf/ieo/pdf/consumption.pdf.

Figure 6.10. World Energy Consumption by Region, 1970–2020.

actions today. They have raised doubts about the accuracy of the model predictions, especially regarding the timing and magnitude of future climate change, as well as the ability of society and ecology to adapt to these changes, and have argued that these uncertainties are too large to warrant redirecting resources away from the needs of the current generation. But the most troublesome obstacle has been the fear that the developed nations will be forced to adopt costly restrictive measures and suffer an economic penalty, while the developing nations, who as noted will account for most of the growth in emissions, will be able to avoid taking action. Many developed nations would like to see the developing world pay for the steps required to reduce greenhouse gas emissions.

For their part, the developing countries, especially newly industrializing countries like China, Brazil, India, and Mexico (as well as Eastern European countries facing severe immediate environmental problems) are unwilling to pay to gain future environmental benefits because of the pressing economic development, social, and environmental problems they confront today. The developing countries would be willing to undertake strategies that minimized their greenhouse gas emissions if the developed nations bore a significant portion of costs required to do so. Not surprisingly, the developed countries resist the suggestion that they should subsidize this process in the developing world.

Because of the relatively backward state of technology and of plant and equipment in much of the developing world, improvements in energy efficiency (in electric power plants, to take one important example) are often more cheaply realized there than in the advanced economies. Thus, a least-cost global strategy for reducing greenhouse gas emissions will entail relatively greater investment in the developing world than in the developed world. The issue is who should pay for this investment. The developing world advocates the transfer of capital from rich to poorer nations for this purpose. The global warming issue has become another vehicle for debate over economic development assistance.

Managing the international dimension of the problem is the most challenging aspect of the global climate change issue. An elaborate international negotiation has gone on for over a decade under the auspices of the United Nations in an effort to reach a consensus on which steps need to be taken and who should bear the costs of agreed reductions. A brief summary of key events follows:

- In 1988, the U.N. Environment Program and the World Meteorological Organization established the Intergovernmental Panel on Climate Change (IPCC), which was charged with providing policymakers with authoritative and up-to-date scientific information.
- In 1990, the IPCC issued its First Assessment Report. The report confirmed that global warming was a threat, and called for global negotiations to address the problem. Later that year the U.N. General Assembly launched negotiations on a Framework Convention on Climate Change. The IPCC's Second Assessment Report, issued in 1996, was a bit more cautious. We have reported the results of the Third Report, issued in 2001, earlier in this chapter.
- In 1992, the Framework Convention on Climate Change was opened for signature at the U.N. Conference on Environment and Development at Rio de Janeiro, the "Earth Summit." The Convention entered into force in 1994, and today 186 nations are parties. The Convention set the ultimate objective of stabilizing atmospheric greenhouse gas concentrations at a level "that would prevent dangerous anthropogenic interference with the climate system." It did not specify what these concentrations should be, nor how to achieve them. It did, however, establish a process for reaching agreement on specific actions. It also assigned most of the responsibility for achieving future reductions in emissions to the industrialized countries. Since the Convention's entry into force, the signatories have met annually in the Conference of the Parties (COP) to monitor progress and continue talks on how best to achieve emission reductions.
- In 1995, at the first COP meeting in Berlin (COP-1), the members, recognizing that the voluntary measures adopted in the Framework Convention would not be sufficient to tackle the problem, began a new round of talks aimed at reaching agreement on mandatory emission reductions by the industrialized countries (the so-called Berlin Mandate.)

Table 6.7. Countries included in Annex B to the Kyoto Protocol and their
emissions targets

	Target (1990 to 2008/2012)
EU-15, Bulgaria, Czech Republic, Estonia, Latvia, Liechtenstein, Lithuania, Monaco, Romania, Slovakia, Slovenia, Switzerland	−8%
United States	−7%
Canada, Hungary, Japan, Poland	−6%
Croatia	−5%
New Zealand, Russian Federation, Ukraine	0
Norway	+1%
Australia	+8%
Iceland	+10%

Source: UNFCCC, *A Guide to the Climate Change Convention Process*, Climate Change Secretariat, Bonn, 2002 (preliminary 2nd edition), at http://unfccc.int/resource/process/guideprocess-p.pdf (visited August 30, 2002).

- In 1997, after more than two years of intense negotiations, the Kyoto Protocol was adopted at COP-3 in Kyoto, Japan.[13]

THE KYOTO PROTOCOL

The 1997 Kyoto Protocol significantly strengthened the Framework Convention by establishing, for the first time, legally-binding limits on the greenhouse gas emissions of individual countries. The limits applied only to the industrialized countries. The developing countries did not agree to any such limits. The individual targets are listed in Annex B of the Protocol, and are reproduced in Table 6.7. The targets cover emissions of the six main greenhouse gases: Carbon dioxide, methane, nitrous oxide, hydrofluorocarbons, perfluorocarbons, and sulfur hexafluoride. The year 1990 was chosen as the baseline for measuring progress on emission reductions.

At Kyoto the United States agreed to reduce its emissions by 7% below 1990 levels, though it has since announced that it does not intend to comply with this commitment (we discuss the U.S. response in more detail later.) The fifteen countries of the European Union (EU) agreed to an 8% overall cut, which will be distributed among the individual nations under a "bubble" scheme that permits some to emit more than others as long as the overall reduction is achieved. For the EU the choice of 1990 as the baseline year made the target easier to achieve because emissions had actually declined between 1990 and

---

[13] Subsequent Conference of Parties have been held:

COP 4 Buenos Aires	November 1998
COP 5 Bonn	November 1999
COP 6 The Hague	November 2000
COP 6-B Bonn	July 2001
COP 7 Marrakesh	November 2001.

1997 following German reunification and the collapse of the East German economy. For similar reasons, the 1990 baseline year did not impose a serious constraint on Russia's future economic growth because of the collapse of the Russian economy after 1990. But for the United States, which enjoyed vigorous economic growth during the 1990s and whose emissions had increased significantly as a result, the reduction target was more challenging.

The Kyoto Protocol also contains a number of other interesting and important provisions that reflect the different circumstances of nation states. First, the measures that countries will take to achieve their agreed reductions are left to them. Second, credit is given for reducing carbon by afforestation, reforestation or other land use practices that provide carbon "sinks." Third, the Protocol encourages cooperation between countries to identify and exploit those opportunities where GHG emissions can be reduced most cost-effectively. For example, coal-burning power plants in developing countries or in the former socialist countries are typically much less efficient than comparable plants in industrialized countries. Rather than spending money on improving the efficiency of its own plants, it will often be much cheaper for, say, the United States to partner with, say, China or Poland to improve the efficiency of plants in those nations. The Kyoto Protocol includes a "Clean Development Mechanism" that allows industrialized nations to obtain credits for carrying out emission reduction projects in the developing countries, and a "Joint Implementation" program to facilitate similar cooperation among the industrialized nations. There is also a provision for emissions trading under which countries can exchange rights to emit GHGs. (The model here is the successful SO_2 emission allowance trading program adopted in the United States during the 1990s and discussed in Chapter 5.) The detailed procedures for implementing these various programs have not yet been finalized.

The Political Response to the Kyoto Protocol in the United States

Negotiating an international treaty like the Kyoto Protocol is evidently very difficult. The benefits are inherently transnational and are not realized by any state unless other states participate. In contrast to military alliance treaties or international arms control agreements, moreover, this type of treaty directly affects the economic growth of nations and hence is of immediate interest to industry, labor, and the general public. Domestic political support is thus vital.

At least in the early stages of the Kyoto negotiations, which were led for the United States by the State Department, American business and labor were not kept informed of progress. Moreover, business and labor were both generally skeptical about the Protocol because the benefits would be collective and distant in time, while the costs of action would be immediate and real. There was also concern about the lack of binding commitments by the developing countries, and the Joint Implementation and Clean Development projects seemed likely to export dollars and jobs overseas. And, as noted earlier, the selection of the 1990 baseline for the targets was disadvantageous to the United States relative to the European Community and the states of the former Soviet

Union. Most importantly, the Administration had little to say about how the United States was going to accomplish the 7% reduction it had agreed to. Would there be new taxes? Would a new emissions trading mechanism be introduced? Would companies that took immediate action to reduce their emissions get future credit? Would the government sponsor research and development programs leading to attractive technologies for future energy efficiency gains? No industry or region knew the extent of the economic burden it would be required to carry. Since no one knew who would be the winners and who would be the losers, each private interest assumed that the worst part of the burden would fall on them. The absence of a clear policy encouraged political opposition. After the Kyoto conference was concluded in 1997, Congress passed a Sense of the Congress resolution that the Protocol should not be ratified unless all countries complied with some restrictions and unless it could be demonstrated that no U.S. jobs would be lost.[14] And the Clinton Administration decided that it would be better not to send the Protocol to the Senate for ratification for the foreseeable future.

In March 2001, President George W. Bush announced that the United States would not ratify the Kyoto protocol, triggering a tremendous outcry from the international community and from environmentalists in the United States. On the one hand, the new President's decision was predictable. United States diplomacy had gotten out in front of the domestic policy process; industry and labor were not informed about the climate issues, and there was little clear information about the government policies that would have to be adopted in order to reach the Kyoto targets. Here was a major shift in economic ground rules with insufficient explanation of the policy objectives and their implementation, and the strong likelihood that important stakeholders would end up as losers. On the other hand, the Bush announcement not only rejected the Kyoto protocol, but also seemed to reject the entire scientific basis of the global warming threat. Nothing was offered in place of Kyoto for dealing with the problem of increasing GHG emissions. In fact, many intermediate options are available between "do nothing" and accepting the Kyoto Protocol in its entirety, but as of this writing the U.S. response to the global warming issue remains unclear.

The Economic Costs of GHG Reductions

The economic cost of compliance is central to the debate over the Kyoto Protocol. Several recent studies have addressed this question. An interesting analysis by Robert Repetto and Duncan Austin of the World Resources Institute explores the dependence of the cost of compliance on different policy assumptions.[15] For example, if a carbon tax is imposed and the resulting revenues are used to finance offsetting cuts in existing taxes that constrain investment, the authors find that a significant portion of the macroeconomic

---

[14] Because the Protocol committed the United States to certain actions by international agreement it required Senate ratification as a treaty.

[15] Robert Repetto and Duncan Austin, "The costs of climate protection: A guide for the perplexed," World Resources Institute, Washington, D.C., 1997.

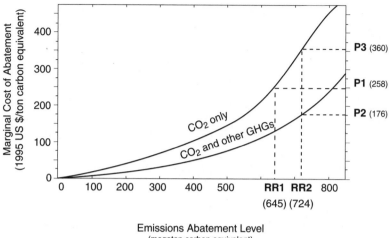

Source: Reprinted from MIT Joint Program on the Science and Policy of Global Change, *Climate Policy Note*, Issue #J-3, April 2000, at http://web.mit.edu/globalchange/www/.

Figure 6.11. The marginal cost of cutting emissions of carbon dioxide and other greenhouse gases in the United States.

burden of the carbon tax can be avoided. The magnitude of the burden will also depend on whether the emission reductions are achieved by the most cost-effective means. Emissions trading and other international joint implementation mechanisms are found to play a key role in achieving cost effective abatement.

Researchers at MIT under the leadership of Professors Henry Jacoby and Ronald Prinn have developed the "MIT Emissions Prediction and Policy Analysis Model."[16] The model's prediction of the marginal costs of GHG abatement in the United States is shown in Figure 6.11. The upper, higher-cost curve describes the marginal cost of abatement of CO_2 alone. The lower curve indicates the marginal cost of abatement and "sink enhancement" strategies for all six classes of greenhouse gases targeted by the Kyoto Protocol. Line RR1 in the figure indicates the required emission reduction if the Kyoto target percentage was applied to CO_2 alone. The total cost of meeting the target in this case is represented by the area under the "CO_2 only" curve up to RR1 (Policy P1). Line RR2 indicates the required reduction if – as is actually the case – all six gases are included in the Kyoto target. The total cost in this case is the area under the "CO_2 plus other GHGs" curve to RR2 (Policy P2). The effect of the Protocol's inclusion of the other greenhouse gases is to increase the required reductions (from 645 to 724 million tons of carbon equivalent), while at the same time increasing the number of cost-reducing opportunities for achieving them. The net effect is to reduce the total annual cost of meeting the target percentage in 2010 from $61 billion for CO_2 alone to $43 billion for

---

[16] See H. D. Jacoby, R. S. Eckaus, A. D. Ellerman, R. G. Prinn, D. M. Reiner, and Z. Yang, "CO_2 emissions limits: Economic adjustments and the distribution of burdens," *The Energy Journal* 18(3,) 31–58 (1997).

all six GHGs. For the industrialized countries as a whole, the effect of the Protocol's multi-gas approach is to reduce the annual abatement cost by 22% from $116 billion to $90 billion.[17]

Another MIT analysis shows that the cost of meeting the Kyoto Protocol targets declines sharply if emission trading is permitted, because of the considerable difference in GHG abatement costs across the world.[18] There is some advantage even if trading is limited to the United States, the European Community, Japan, and the countries of Eastern Europe and the former Soviet Union. But if trading is also allowed with the developing countries the savings are much larger, because the cost of GHG abatement is often so much lower in the developing world.

It is important to appreciate the analytic underpinning required to make estimates of this kind. The MIT model includes a model of the global economy that comprises an interconnected set of national and regional entities. Each have supplies of capital, labor, and land inputs as well as consumer demand functions and production technologies, and each is divided into several production and consumption sectors. The economic model solves for a set of product and factor input prices that balance supplies and demands in each market in each region for each time period. The model also computes the anthropogenic emissions of significant greenhouse gases that are associated with this level of economic activity. The economic model is linked to a coupled model of atmospheric chemistry and climate.

This system of models is first used to calculate the GHG emissions associated with a "business as usual" scenario of economic activity, and then to relate these emissions to changes in atmospheric concentrations of GHGs and ultimately to the resulting climate changes. It must then be able to relate a change in policy – such as the imposition of a carbon tax – to changes, first, in the level of economic activity, then in the associated atmospheric emissions, then in the atmospheric concentrations, and then in atmospheric temperature. All of this requires a remarkable blending of natural science and social science modeling – a very valuable, but quite rare skill.

In 1997, the MIT group used its models to estimate the economic adjustment cost of reducing global fossil-fuel-related CO_2 emissions to a rate 20% below the business-as-usual expectation by the year 2100. In this modeling exercise, it was assumed that all of the burden of emission reductions would fall on the OECD countries, and that CO_2 emissions from the non-OECD countries would not be constrained. Today, world emissions of carbon dioxide from the use of fossil fuels are a little over 6 gigatonnes per year of carbon equivalent. In the business-as-usual scenario, these emissions were projected to rise to 19 gigatonnes in the year 2100. The MIT researchers then calculated the economic cost of reducing fossil emissions to 14 gigatonnes by imposing a system

[17] Policy P3 in Figure 6.11 shows the cost of meeting the Kyoto multi-gas target with CO_2 abatement strategies alone. For the industrialized countries as a whole, the total annual cost would be $174 billion, almost twice as much as if reductions in all gases are pursued.

[18] A. D. Ellerman and A. Decaux, "Analysis of post-Kyoto emissions trading using marginal abatement curves," Report #40, Joint Program on the Science and Policy of Global Change, MIT, October 1998.

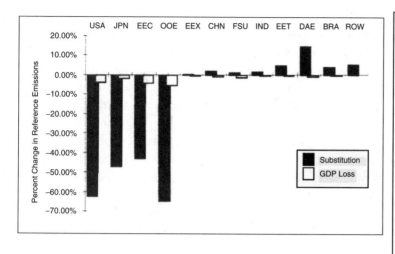

Source: Reprinted from H. D. Jacoby et al, "CO_2 emissions limits: Economic adjustments and the distribution of burdens," *The Energy Journal* 18 (3), (1997), 44.

Figure 6.12. Percentage change in carbon emissions between 2000 and 2100 by region, assuming a global emission reduction of 20% below the 'business-as-usual' scenario, with emission restrictions applied only to the OECD countries.

of emissions quotas on the OECD countries. (The calculation assumed the availability of a backstop noncarbon fuel electricity source at a cost of $.15 per kwhr.)

The predicted economic cost, calculated as the present (that is, discounted) value of consumption foregone between 2000 and 2100, and expressed as a percentage of all future consumption during this period, was found to be 1.5% for the United States, 2.3% for the European Community, and 0.5% or less for India and China. Although these percentages are small, the absolute quantities of foregone consumption are staggeringly large.

The international distribution of emission reductions is indicated in Figure 6.12. In order to achieve the global target of 20%, carbon emissions in the industrialized countries are projected to decline by about 50% relative to the business-as-usual scenario and to remain roughly unchanged or even to increase slightly in the rest of the world. The climate model predicts that the effect of the emission reduction will be to reduce the atmospheric concentration of CO_2 in the year 2100 from 741 ppm (in the business-as-usual scenario) to 629 ppm – still more than twice the pre-industrial level. The predicted mean global temperature increase in the year 2100 is reduced from +2.37°C to +2.00°C.[19]

What is startling about this analysis is how small the reduction in the atmospheric concentration of carbon dioxide is in relation to the business-as-usual scenario, even when relatively severe measures are adopted and there is compliance by all relevant

[19] It should be noted that the economic impact of the 20% reduction estimated by the MIT group did not take account of any economic gains from the avoided climate change.

countries (in this case all the OECD countries.) Results such as these have led some to ask whether more aggressive technology strategies for improving energy productivity and reducing carbon intensity could be adopted, or whether cost-competitive carbon sequestration technologies might be available. For it is clear that if the actual climate impacts of GHG emissions turn out to be on the more pessimistic side of the predicted range of possibilities, the measures that have been considered in most of the studies to date will prove to be wholly inadequate.

The conclusion to this chapter is that the global climate change problem will be with the United States and the world for many decades to come. The scientific issues presented by global warming are enormously complex, and the economic and political issues involved in developing a policy response that must be global in extent will be extraordinarily difficult for governments to manage. Many readers of this book will undoubtedly be involved with one or another aspect of global warming in their future careers.

7

Nuclear Power and Its Fuel Cycle

No technological system more dramatically illustrates the central themes of this book – the complexity of real world applications of technology and the pitfalls of ignoring the social, political, and environmental dimensions of innovation – than nuclear power. Once widely seen as an energy source of almost unlimited potential, nuclear power is today expanding in just a handful of countries. In most countries with operating nuclear power stations there are no plans to build additional nuclear plants, and some countries have made formal decisions to phase out their existing reactors as quickly as possible.

Despite its limited growth prospects, nuclear power is today playing an important role around the world with nearly 440 plants supplying 17% of the world's electricity. In some countries, the level of dependence is much higher. France derives 76% of its electricity from nuclear power, and other heavily nuclear-reliant countries include Belgium, Japan, and South Korea (see Table 7.1). The world's largest nuclear power program is in the United States, where more than 100 plants provide 20% of the nation's electricity. Keeping these plants operating safely, reliably, and economically is a vital task for private firms and governments around the world. But with few new nuclear plants being built, almost every energy forecast projects a gradual decline in the nuclear share of world electricity supplies. In the United States, for example, where the last time construction began on a new nuclear plant was more than 20 years ago, the government projects that nuclear generation will decline 7% from its present level by the year 2020.

There is no single, simple explanation for the troubled state of the nuclear power industry. One important cause is the extraordinary difficulty that almost every country has had in finding an acceptable solution to the problem of nuclear waste disposal. Another is the continuing concern over the safety of nuclear power reactors and other nuclear facilities, despite an industry safety record that – with the dramatic exception of the Chernobyl disaster in the Ukraine in 1986 – has been impressive when compared with the available alternatives. A third factor is the risk that nuclear power will contribute to the further spread of nuclear or radiological weapons to governments or to terrorist groups. Each of these issues has negatively influenced public opinion toward nuclear technology. But probably the biggest obstacle to further nuclear development today is that in most parts of the world new nuclear power plants are too expensive to compete with other supply options. The economic problem is indicated in Figure 7.1, which

Table 7.1. Countries with highest dependence on nuclear power (2000)

Country	Number of plants	Percentage of electricity generation from nuclear
France	59	76.4
Lithuania	2	73.7
Belgium	7	56.8
Slovakia	6	53.4
Ukraine	13	47.3
Bulgaria	6	45.0
South Korea	16	40.7
Hungary	4	40.6
Sweden	11	39.0
Switzerland	5	38.2
Slovenia	1	37.4
Japan	53	33.8
Finland	4	32.1
Germany	19	30.6
Spain	9	27.6
United Kingdom	35	21.9
Czech Republic	5	20.1
United States	104	19.8
Russian Federation	23	14.9
Canada	14	11.8
WORLD TOTAL	438	17.0

Source: International Atomic Energy Agency

shows a recent U.S. projection of levelized electricity generation costs from new nuclear plants and alternative sources.[1]

It is possible that all these problems will eventually be overcome. Indeed, overcoming them may well turn out to be essential if the world is to achieve significant reductions in greenhouse gas emissions from fossil fuel combustion, as discussed in Chapter 6. Yet the problems facing nuclear power are formidable, and at this juncture it is not at all clear that they can be resolved successfully. Part of the difficulty is that they expose some of the deepest fault lines in the debate about how to control technology in democratic societies. How should authority to regulate nuclear technology be distributed among local and central jurisdictions – an issue which in the United States has often pitted local, state, and federal authorities against each other? What are the rights of activist minorities vis-à-vis an ambivalent, disengaged majority? What is the appropriate role for technical experts in the resolution of issues in which there is a strong political component? And when technologies raise major issues of public risk, to what

---

[1] As Figure 7.1 shows, the dominant contributor to the total cost of nuclear electricity is the power plant construction cost. The fuel and operating and maintenance costs are relatively low, which explains why existing nuclear power plants whose capital costs have been amortized are highly competitive with most alternative sources of electricity.

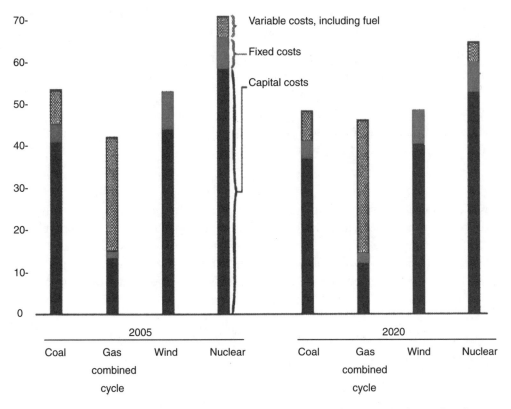

Source: Reprinted from U.S. Energy Information Administration, *Annual Energy Outlook* 2002, Reference Case Forecast at http://www.eia.doe.gov/oiaf/aeo/ (December 2001).

Figure 7.1. Projected levelized electricity generation costs, 2005 and 2020 (in mills per kilowatt hour ($2000)).

extent should society rely on private firms competing in the marketplace to manage them?

Whatever the outcome of the debate about nuclear power, it is important to understand the circumstances that led to the present state of near-paralysis in the industry. In the limited space available here we cannot hope to give a full accounting of the history of this troubled technology. In this chapter, we review some key technical features of nuclear power reactors and fuel cycles. In the next two chapters, we will focus on two issues of great importance to the future of nuclear power, both of which highlight the inseparability of technical and institutional considerations: Nuclear waste management and disposal, and the proliferation of nuclear weapons.

Nuclear Physics Background

We begin by briefly describing the two most important nuclear phenomena in the nuclear power fuel cycle: Radioactive decay and nuclear fission.

Radioactive Decay

The three types of spontaneous radioactive decay processes that are of primary interest in nuclear energy applications are alpha, beta, and gamma decay.

Alpha Decay. In the alpha decay process the decaying isotope – the parent nucleus – emits a helium nucleus (the alpha particle) as it decays to the daughter species. The alpha decay reaction can be written:

$$^{M}_{Z}X \xrightarrow{\alpha} {}^{M-4}_{Z-2}Y + {}^{4}_{2}He.$$

Notice in this equation the convention of placing the atomic number (Z), the number of protons in the nucleus, to the lower left of the chemical symbol, and the mass number (M), the total number of protons and neutrons, to the upper left. It is a characteristic of all nuclear reactions occurring in nuclear reactors that: (a) the total number of protons and neutrons is conserved, and (b) the charge is also conserved. Thus, for the general nuclear reaction:

$$^{A_1}_{Z_1}P + {}^{A_2}_{Z_2}Q \rightarrow {}^{A_3}_{Z_3}R + {}^{A_4}_{Z_4}S + {}^{A_5}_{Z_5}T,$$

it is necessary that:

$$A_1 + A_2 = A_3 + A_4 + A_5$$

and

$$Z_1 + Z_2 = Z_3 + Z_4 + Z_5.$$

Most alpha-emitting nuclei are found among the heavier isotopes at the upper end of the periodic table – the so-called actinides, which follow the element actinium (atomic number 89). For example, the two naturally occurring isotopes of uranium, uranium-235 and uranium-238, are both alpha-emitters:

$$^{235}_{92}U \rightarrow {}^{231}_{90}Th + \alpha$$
$$^{238}_{92}U \rightarrow {}^{234}_{90}Th + \alpha.$$

Alpha particles give up their kinetic energy by causing ionization in the matter through which they pass. Most of the energy is deposited close to the site of the alpha-emitting nucleus. The range of alpha particles is just a few centimeters in air, and is much shorter in solid media. (Most alpha particles are incapable of penetrating a sheet of paper.) For this reason, alpha-emitting radionuclides are not an external hazard to humans. Even if they reside on the surface of the skin, the alpha radiation cannot penetrate the skin's outer layers. However, if the alpha emitters are ingested or inhaled they can be extremely hazardous. Great damage can be done to the tissue close to the decaying nuclide, because almost all of the alpha energy is deposited within a very small range of the source.

Beta Decay. The beta decay reaction can be thought of as the transformation of a neutron in the nucleus of an isotope into a proton, followed by the emission of a negatively charged electron to maintain the charge balance. The atomic number of the isotope increases by one:

$$^M_Z X \xrightarrow{\beta} {}^M_{Z+1}Y + {}^0_{-1}e.$$

(The mass of the electron is ignored in this equation because it is so small, about 1/1840 the mass of a proton.) Beta particles also lose kinetic energy by ionization. The range is somewhat longer than that of alphas, and externally located beta emitters can cause skin burns. Beta emitters are also dangerous when ingested or inhaled.

Gamma Decay. In gamma decay a high-energy photon of electro-magnetic energy is emitted from an excited nucleus as it undergoes a transition to a lower energy state. Gamma radiation is similar to x-radiation, but the photons are an order of magnitude more energetic on average (although the energy ranges overlap to some extent.) The range of gamma photons is much longer than alpha or beta particles. A beam of gamma photons passing through air travels about 100 meters before it is attenuated to half its original strength. In water the 50% attenuation distance is on the order of tens of centimeters. External exposure to gamma emitters can be extremely hazardous to humans.

Radioactive Decay Law

The decay of any radioactive isotope can be described by the first-order reaction rate expression:

$$\frac{dN(t)}{dt} = -\lambda N(t), \tag{1}$$

where $N(t)$ is the number of nuclei of the isotope present at time t and λ is the decay constant for the isotope, with units of reciprocal time. The solution to this equation is

$$N(t) = N(0)e^{-\lambda t}, \tag{2}$$

where $N(0)$ is the number of nuclei present at time 0. The *half-life* of the isotope, $t_{1/2}$, defined as the time required for half of the nuclei originally present to decay away, is given by

$$N(t_{1/2}) = N(0)/2 = N(0)\exp(-\lambda t_{1/2}).$$

Thus

$$t_{1/2} = (1/\lambda)\ln 2. \tag{3}$$

Some radioactive nuclei have extremely short half-lives, on the order of seconds or even less. At the other end of the scale, some isotopes are only slightly unstable. For example, the half-lives for alpha decay of ^{235}U and ^{238}U are 710 million years and 4.5 billion years, respectively.

The roughly six times faster decay rate of ^{235}U means that the ^{235}U:^{238}U ratio in uranium ore deposits occurring in nature was significantly higher earlier in the Earth's history than it is today – high enough, in fact, to give rise to the phenomenon of "natural" nuclear reactors. Unlike its more plentiful sister isotope, uranium-235 can be fissioned by low-energy neutrons, and while natural uranium ore today contains only 0.711% of uranium-235 – not enough to sustain a fission chain reaction – in the remote past the enrichment was high enough for a chain reaction to be sustainable under the right conditions. In fact, several such naturally occurring nuclear reactors are known to have operated in uranium ore deposits at Oklo in Gabon about 1.7 billion years ago. From equations (2) and (3) displayed earlier, we can calculate that the proportion of uranium-235 at that time was:

$$\frac{0.711e^{+(\ln 2/0.71\times10^9)(1.7\times10^9)}}{99.289e^{+(\ln 2/4.5\times10^9)(1.7\times10^9)}} = 0.029$$

or almost 3% – roughly what it is in modern light water reactors. (The first evidence for the existence of these natural reactors emerged when measurements conducted by French scientists on uranium ore obtained from deposits in Gabon revealed an abnormally low isotopic concentration of ^{235}U, that is, well below 0.711%. The explanation was that some of the original ^{235}U had been consumed during operation of the natural reactor long ago.)

THE FISSION PROCESS

The most important nuclear process occurring in nuclear reactors is neutron-induced fission. In the fission reaction, a neutron is absorbed by the fissionable target nucleus. In most power reactors this is usually either uranium-235 or plutonium-239. The initial product of the absorption is an excited compound nucleus. The compound nucleus then either undergoes a transition to its ground state or, with higher probability, splits into two smaller nuclei (the fission products). The fission event results in the release of about 200 million electron volts (MeV) of energy (equal to about 3.2×10^{-9} joules). Two or three energetic neutrons are also released in the fission reaction (the average per fission event is about 2.5), along with several gamma photons and beta particles (energetic electrons). It is the release of more than one neutron in each fission event (neutron "multiplication") that makes a sustainable fission chain reaction possible.

The reaction scheme for uranium-235 can be written as:

$$^{235}_{92}\text{U} + \text{n} \rightarrow {}^{236}_{92}\text{U}^* \rightarrow \text{fission products} + 2 \text{ or } 3 \text{ neutrons} + \gamma\text{s} + \beta\text{s}$$
$$\searrow$$
$$^{236}_{92}\text{U},$$

where the asterisk denotes an excited nuclear state. The relative probabilities of fission and neutron capture depend on the energy of the incoming neutron. For low-energy

Table 7.2. Energy distribution in fission reactions

	MeV/fission
Kinetic energy of fission fragments	165 ± 5
Kinetic energy of neutrons	5
Prompt gamma photons	7 ± 1
Delayed gammas from fission fragment decay	7 ± 1
Delayed betas from fission fragment decay	8 ± 1
Total recoverable energy	192
Neutrino energy (unrecoverable)	(11)

(thermal) neutrons, the probability that fission will occur following a neutron capture is greater than 80%.

There is usually a significant difference in the mass numbers of the two fission fragments produced in a fission reaction. Many possible combinations of fission products may be formed. Altogether, more than 300 different fission products have been identified. Even the most frequently produced of these appear only a small percentage of the time.

An energy release of 200 MeV per fission event means that the complete fissioning of 1 g of U-235 would generate almost 1 megawatt day of energy. This is about 2 million times the energy released in the combustion of a gram of coal. Alternatively, we can say that the energy content of 1 g of U-235 is equivalent to that of about three tons of coal or nearly 14 barrels of oil.

Most of the energy unleashed in the fission reaction is released almost instantaneously. About 165 MeV is taken up in the kinetic energy of the fission fragments (see Table 7.2). Essentially all of this energy is converted to thermal energy via Coulomb interactions between the fission products and neighboring charged nuclei which take place within about a micron (10^{-4} cm) of the site where the original fission took place. The kinetic energy of the two to three neutrons accounts for another 5 MeV, and this too is converted into heat as the neutrons collide with the nuclei in the reactor, thereby increasing their average kinetic energy. The gamma photons released at the time of the fission event (the "prompt" gammas) contribute about 7 MeV more.

However, about 8% of the fission energy (roughly 15 MeV) is released only gradually as the radioactive fission products decay, emitting gamma and beta radiation as they undergo transitions to stable, nonradioactive isotopes. The half-lives of these fission products range from fractions of a second to millions of years, so that beta and gamma emissions continue for long after the nuclear chain reaction has stopped. In the period immediately following the shutdown of a reactor, large amounts of this decay heat continue to be generated, and it is of the utmost importance to maintain uninterrupted cooling of the reactor in the hours and days after shutdown to prevent the fuel from overheating and possibly melting. Over longer time scales the thermal power diminishes, but the biological risks from the beta and gamma-emitting fission products in the fuel (as well as the alpha-emitting actinides) remain considerable. Protecting human populations and the environment from the effects of these long-lived radioactive species is the primary goal of radioactive waste management, as we will discuss in the next chapter.

Fissile and Fertile Isotopes. Only a few isotopes can be fissioned by neutrons travelling at slow speed. Of these so-called *fissile* isotopes, the only naturally occurring one is uranium-235. Man-made fissile isotopes include plutonium-239 and uranium-233. Plutonium-239 is produced via a neutron capture reaction with a uranium-238 nucleus. The compound uranium-239 nucleus quickly undergoes two successive beta decay reactions to yield plutonium-239:

$$^{238}_{92}\text{U} + ^{1}_{0}\text{n} \rightarrow ^{239}_{92}\text{U}^* \xrightarrow{t_{1/2}\,=\,23.5\text{ min}} ^{239}_{93}\text{Np} + ^{0}_{-1}\text{e} \xrightarrow{t_{1/2}\,=\,2.35\text{ days}} ^{239}_{94}\text{Pu} + ^{0}_{-1}\text{e}.$$

Uranium-233 is produced through a neutron-capture reaction in thorium-232 (the only naturally occurring thorium isotope.)

$$^{232}_{90}\text{Th} + ^{1}_{0}\text{n} \rightarrow ^{233}_{90}\text{Th} \xrightarrow{t_{1/2}\,=\,22.2\text{ min}} ^{233}_{91}\text{Pa} + ^{0}_{-1}\text{e}$$
$$\downarrow t_{1/2}\,=\,27\text{ days}$$
$$^{233}_{92}\text{U} + ^{0}_{-1}\text{e}.$$

Though not themselves fissile, uranium-238 and thorium-232 have special significance as the only naturally occurring species from which fissile isotopes can be produced. They are called *fertile* isotopes. Finally, many heavy isotopes (including fertile uranium-238 and thorium-232) can be made to undergo fission if the energy of the impinging neutron is high enough (i.e., if the neutrons are "fast"); these are referred to as *fissionable* isotopes.

Physics of Nuclear Reactors

The condition of the neutron population in a reactor at any given instant is described by the following neutron balance:

Net rate of neutron accumulation = rate of production − rate of destruction.

The neutrons are produced in fission reactions and are removed by absorption in fissile and fertile nuclei, parasitic capture in structural or other materials, and leakage out of the reactor core. After absorbing a neutron, some of the fissile nuclei undergo fission, giving birth to a new generation of neutrons, and thus the chain reaction continues. When the rates of neutron production and destruction are exactly in balance, the reactor is said to be critical. The quantity used to characterize the degree of criticality of a reactor is the reactivity, ρ, defined conceptually as the ratio of the net to the total rate of neutron production, that is,

$$\rho = \frac{\text{neutron production rate due to fission} - \text{neutron destruction rate}}{\text{neutron production rate due to fission}}.$$

When the reactivity is positive ($\rho > 0$) there is net production of neutrons, and the reactor is said to be super-critical. When the reactivity is negative ($\rho < 0$), there is net neutron destruction, and the reactor is sub-critical.

Reactor operators can control the reactivity of a reactor by adjusting the position of the control rods. These are structures containing neutron "poison" – material with a high affinity for neutron capture, such as the boron isotope ^{10}B. By inserting the control rods into the core, the operators can reduce the net production rate (and hence the reactivity). Conversely, removing the rods increases the reactivity.

To illustrate the sensitivity of reactor control, suppose the rods are withdrawn from an exactly critical reactor to a position at which the reactivity is equal to +0.001. Since the reactivity is positive, the neutron population will start to rise. A reactivity of 0.001 means that each "generation" of neutrons will be 0.1% larger than its predecessor. The initial rate of increase is extremely rapid. The average lifetime of neutrons in a typical power reactor is 10^{-4} seconds, which means that in the course of one second 10,000 generations of neutrons are born and destroyed. So, after one second the neutron population should have increased by a factor of

$$1.001^{10,000} \cong 22,000.$$

Because the fission rate, and therefore the reactor power, increase proportionally with the neutron population, this simple calculation suggests that even the slightest perturbation to the neutron balance in the direction of positive reactivity ought to cause the reactor quickly to blow itself apart. Fortunately this does not happen. Two offsetting factors ensure that reactors are not inherently unstable. First, a small fraction (less than 1%) of the two or three neutrons that are emitted in each fission event do not appear instantaneously, but rather are released as certain fission products decay. The average half-life of these fission products is about 12 seconds, and the so-called *delayed neutrons* provide a crucial margin of safety. As long as the positive reactivity – the fractional neutron surplus produced per generation – is less than the delayed neutron fraction, it is the delayed neutrons that play the decisive role in sustaining the chain reaction. Under these conditions the neutron population increases at a rate determined by the rate of release of the delayed neutrons. The neutron doubling time is measured in seconds, rather than in thousandths of a second. With the buildup of neutrons and power occurring on a timescale of seconds, the control rods can be inserted into the reactor quickly enough to offset the reactivity increase that caused it, absorb the excess neutrons, and shut down the chain reaction. (Of course, if the addition of reactivity were larger than the delayed neutron fraction, the delayed neutrons would no longer regulate the rate of neutron buildup. The prompt neutrons alone would multiply from one generation to the next. This would occur extremely rapidly, and a nuclear "runaway" would result. The reactor might indeed then blow itself apart.)

The other important contributors to reactor stability are a series of negative feedback effects that automatically cause the reactivity to decline as the neutron population level rises. In most reactors, the neutron balance is temperature-sensitive. As the temperature of the core rises, relatively more of the neutrons are captured in nonfissile nuclei.[2] The effect is to shut the chain reaction down.

Nuclear Reactor Types

Reactors can be classified according to whether the neutron population is "fast" or "thermal." This refers to the average energy of the neutrons in the core. Fission neutrons are born with an energy of 2–3 MeV. In a thermal reactor, most of the fission neutrons are slowed down in a series of collision (scattering) reactions before being absorbed, and the neutron population is in approximate thermal equilibrium with the reactor materials. In a fast reactor, the neutrons undergo far fewer collisions before being absorbed, and the average energy of the neutron population is much higher. The main motivation for thermal reactors is that a thermal neutron passing through fuel has a much higher probability of causing a fissile isotope to undergo fission per unit of distance traveled than a fast neutron. Because the "critical mass" of fissile material – the minimum amount required to sustain a chain reaction – is roughly inversely proportional to this probability, the critical mass of both uranium-235 and plutonium-239 is smaller in thermal reactors than in fast reactors. (The main advantage of fast reactors, as we shall see below, is that they are better able to sustain nuclear breeding.)

The material used to slow down neutrons in thermal reactors is called the *moderator*. Neutrons lose energy most effectively in collisions with light nuclei. A good moderator is thus one that contains isotopes with low atomic weight and that also have a low affinity for neutron capture. Good candidates are hydrogen, deuterium (the hydrogen isotope that contains a neutron as well as a proton in the nucleus and thus has an atomic weight of 2), beryllium (atomic weight 9), or carbon (atomic weight 12). The moderators in most common use in power reactors are light water (H_2O), heavy water (D_2O), and graphite.

In light water reactors (LWRs), which are the most common type of power reactor in use today, the same material, H_2O, is used as both moderator and coolant. In fast reactors, water cannot be used as a coolant because the hydrogen is too effective as a neutron moderator. The most common coolant in fast reactors is liquid sodium (atomic weight 23).

Another important characteristic of reactors has to do with the rate at which fissile nuclei are replaced in the core. Most types of nuclear reactors contain large quantities of fertile isotopes, and some of the fission neutrons undergo neutron capture reactions in these fertile nuclei to produce new fissile nuclei. If the number of neutrons created per neutron absorbed in fissile material is greater than two, it is theoretically possible to create fissile nuclei at a faster rate than they are being consumed.[3] One of the newly

---

[2] An important contribution is provided by the nonfissile U-238 nuclei, which absorb more effectively at higher temperatures because of Doppler broadening of the neutron capture resonances.

[3] The number of fission neutrons produced per neutron *absorbed* in fissile material is lower than the number of neutrons produced per fission event, because some of the absorbed neutrons, instead of inducing fission,

Table 7.3. Nuclear reactor types

	Light water reactors (LWR)				Liquid metal cooled fast breeder reactor (LMFBR)
	Pressurized-water reactor (PWR)	Boiling-water reactor (BWR)	Heavy water reactor (CANDU)	High temperature gas reactor (HTGR)	
Fissile isotope (% enrichment)	^{235}U ($\sim$3%)	^{235}U ($\sim$3%)	^{235}U (0.711%)	^{235}U ($\geq$10%)	^{239}Pu ($\geq$10%)
Fertile isotope	^{238}U	^{238}U	^{238}U	^{232}Th	^{238}U
Coolant	H_2O	H_2O	D_2O	Helium	Liquid Na
Moderator	H_2O	H_2O	D_2O	Graphite	None
Conversion ratio	$\sim$0.55	$\sim$0.55	$\sim$0.7	$\sim$0.7	$\sim$1.2

created neutrons would cause another fission to keep the chain reaction going, while the second would be captured in a fertile isotope, forming a fissile nucleus that would replace the one consumed in the fission reaction. Any additional neutrons would be available to produce more of the fissile isotope. In practice, however, this is a very difficult feat to accomplish, because even under the most favorable circumstances the number of neutrons produced per neutron absorbed in fissile nuclides is barely more than two, and some neutrons are inevitably lost to nonproductive absorptions in structural and other materials and to leakage out of the reactor.

The ratio of the production rate of fissile material to the rate at which it is consumed is called the conversion ratio. Reactors in which the conversion ratio is less than one are called converters. Reactors in which the conversion ratio is greater than one are called breeders. In thermal reactors, breeding is only possible with uranium-233. In fast reactors, where there is less opportunity for neutrons to be captured nonproductively as they slow down, both uranium-233 and plutonium-239 are capable of sustaining breeding.

From this brief description, we see that reactors can be classified according to (a) whether the neutron spectrum is fast or thermal, (b) whether the reactor is a breeder or a converter, (c) the primary fissile isotope, (d) the primary fertile isotope, (e) the type of moderator (for thermal reactors), and (f) the coolant type. The characteristics of the main power reactor types in commercial operation or under development are summarized in Table 7.3.

More than 90% of the power reactors in use around the world today are light water reactors. In LWRs, the uranium fuel is enriched in the fissile isotope uranium-235 from its natural concentration of 0.711% up to about 3%. The enrichment is necessary to offset the adverse effect on the neutron balance of neutron capture reactions in the light water coolant/moderator. In heavy water, the probability of neutron capture is

are captured to form the higher isotope (e.g., uranium-236 from uranium-235, and plutonium-240 from plutonium-239).

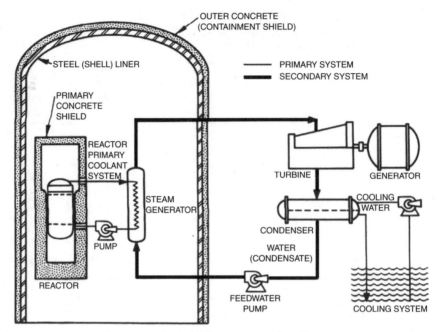

Figure 7.2. Pressurized water reactor flow diagram.

significantly lower, and heavy water reactors of the CANDU type developed originally in Canada (CANDU stands for Canadian Deuterium-Uranium) can use natural (that is, unenriched) uranium fuel.

About two-thirds of all light water reactors are of the pressurized water reactor (PWR) type. In PWRs the water is maintained at a high enough pressure to prevent boiling even at the maximum temperature reached in the reactor core, which is about 300°C. The reactor flowsheet is shown schematically in Figure 7.2. The hot coolant flows from the reactor to a steam generator, where it boils lower pressure water on the secondary side. The steam in the secondary loop is used to drive a turbine-generator. The primary coolant is recirculated to the reactor, and the steam, after passing through the turbine, is condensed and returned to the steam generator. The reactor vessel and the other components of the primary loop are contained in a massive concrete containment vessel.

In the other type of LWR, the boiling water reactor (BWR), the water coolant/ moderator is at a lower pressure and is partially converted to steam in the core. The steam is separated from the water as it leaves the reactor and is sent directly to drive the turbine. The water is returned to the reactor as is the steam after it leaves the turbine.

There are many other possible combinations of coolant, moderator, and fuel type. Another kind of reactor, which became notorious following the devastating accident at Chernobyl in 1986, is the RBMK – a graphite-moderated, light-water cooled system that was developed and introduced on a commercial scale in the Soviet Union. We discuss the RBMK reactor and the Chernobyl accident in more detail in subsequent paragraphs.

Nuclear Reactor Safety

There are two overriding safety requirements in nuclear reactors. The first is to ensure that the nuclear chain reaction can be stopped dependably and rapidly under all possible operating conditions. This is typically achieved through a combination of control rod insertions and the negative reactivity feedback mechanisms operating in the fuel and other core materials. The second requirement is to ensure the removal of the fission product decay heat, which continues to be generated in large amounts even after the chain reaction has been stopped. Immediately after shutdown, the decay power is equal to several percent of the full power level. For a large power reactor generating 1,000 megawatts of electricity in normal operation, this means that immediately after shutdown the core continues to produce about 200 megawatts of decay heat. The decay power declines over time, but in the hours and days after shutdown the rate of decline is quite slow. Thus, a full hour after shutdown the core is still producing nearly 100 megawatts of power, and even after a day it is generating 20 megawatts or so.

Failure to remove this heat promptly may result in fuel overheating and melting, followed by the energetic release of radionuclides. In the worst case, there is the possibility that the fuel will melt its way through the vessel containing the core, through any additional containment structures, and out into the environment. Even without a full core meltdown severe problems can arise. Much of the radioactive inventory in the core consists of volatile fission product isotopes like iodine-131, cesium-134, cesium-137, and the noble gases, and rupturing of the cladding can result in their release. In reactors in which water is present, there is also the possibility of steam explosions with enough over-pressure generated to breach the reactor vessel. It is also possible that large amounts of hydrogen will be formed in chemical reactions between the steam and reactor materials, creating the risk of containment-breaching explosions.

All of this means that reliable systems for removing the post-shutdown decay heat are a crucial safety feature of nuclear power reactors. Such systems can be either "passive," that is, they rely only on natural heat removal mechanisms, or "active," requiring forced cooling and, possibly, the intervention of operators to initiate or control the cooling process. In theory, exclusive reliance on passive systems would be preferable. There would then be no need to plan for the possibility that pumps would fail to come on, valves would fail to open, operators would fail to react promptly and accurately, and so on. On the other hand, there are strong economic incentives to build reactors with high power density (that is, high power output per unit volume of core), since compact reactors are generally less expensive. But the higher the power density, the more likely it is that purely passive systems for decay heat removal will be insufficient and that active cooling will also be required. In commercial LWRs (which are technological descendants of submarine propulsion reactors, where compactness is imperative), heavy reliance is placed on active engineered systems to ensure that high-pressure cooling water can be forced through the core in emergency situations.

In order to achieve the necessary level of reliability, much redundancy is built in to the reactor design, including conservatively designed pumps, pipes, and other feedwater system components; backup cooling systems with independent power supplies to drive them; continuous monitoring and periodic inspection of safety systems to ensure their constant availability; fault-tolerant computer controls; and a highly trained operating staff following precise and well-rehearsed procedures.

If all else fails, there is a massive, reinforced-concrete containment structure that encloses the reactor, the primary coolant loop, and some of the ancillary systems. This "defense-in-depth" safety strategy has worked well for the more than 400 commercial LWRs in service around the world. The most serious LWR accident on record occurred twenty years ago, at the Three Mile Island nuclear plant in Pennsylvania. A combination of equipment failures and human error led to the destruction of the core at Three Mile Island Unit 2, with financial losses running into the billions of dollars. But there were no injuries, and radiation doses to the external population rose only slightly above natural background levels. There have been near-misses at other plants, but on the whole the LWR fleet has compiled an impressive safety record with no fatalities despite several thousand reactor-years of operation.

Effective performance in the past, while a good indicator of future safety, is of course no guarantee of it, and it is important to regulators, to plant owners, and to the general public to have as clear a picture as possible of whether an accident is likely to occur in the future, and to know what can be done to reduce this risk. One of the characteristics of systems like nuclear power reactors is that they (fortunately) fail so rarely that previous operating experience is of only limited value for future risk estimation. Moreover, the consequences of even a planned failure would be so costly that another possible approach to risk assessment, deliberately stressing the system until it fails, is simply not practical in this case.

Sophisticated probabilistic risk analysis (PRA) methods have therefore been developed to assess safety risks. These techniques involve gathering empirical data about the failure behavior of individual system components, and then constructing hypothetical scenarios about how individual failures might combine to cause a general system failure. Initially developed for space flight systems, PRA methods were first applied to nuclear reactors in the mid-1970s by a team led by MIT nuclear engineering professor Norman C. Rasmussen. These early calculations turned out to be quite controversial, and their use and misuse by parties on both sides of the nuclear safety debate is a classic case study of what can happen when scientific and technical knowledge is introduced into a highly politicized arena. Since then, however, the methods pioneered by Rasmussen and his colleagues have come to be widely and routinely used throughout the nuclear power industry and have also been applied to a variety of other large-scale, complex technologies, including aircraft and chemical processing plants. Chapter 11 discusses their application to liquefied natural gas facilities.

THE ACCIDENT AT CHERNOBYL

The world's worst nuclear accident occurred in 1986 at Reactor Number 4 of the four-unit Chernobyl nuclear power station in the Ukraine, then part of the Soviet Union. The accident destroyed the reactor and surrounding buildings, killing more than 30 people in the days and weeks that followed, badly contaminating millions of acres of agricultural land, forcing the evacuation and permanent resettlement of about 130,000 people, and depositing radioactive fallout across vast areas of the Soviet Union, Scandinavia, central and southern Europe, and elsewhere. Nearly every country in the northern hemisphere measured some elevation in radiation levels. As a result of the accident, more than half a million people were exposed to significant doses of radiation and will be monitored for health consequences for the rest of their lives. The number of premature deaths associated with the Chernobyl disaster is expected to run into the tens of thousands.

The Chernobyl unit was a 1,000 MWe reactor of the RBMK type, a boiling-water-cooled, graphite-moderated design which at the time of the accident was the workhorse of the Soviet nuclear power program. In RBMK reactors, the uranium fuel, enriched to about 2% in ^{235}U, is contained in individual pressure tubes about 10 meters in length through which cooling water flows and is converted to steam by the heat from the fuel. The steam is piped directly to a turbine-generator. In the Chernobyl reactor, there were about 1,700 of these fuel channels, which were separated by columns of the graphite moderator.

The accident occurred during a test of a safety system. In the course of the test, which was poorly planned and should not have been conducted with the reactor in its then-operating condition, the flow of cooling water through the core was interrupted. Large steam bubbles or "voids" began to form in the core. At the low power level at which the reactor was operating, there was a "positive void coefficient" of reactivity, meaning that an increase in the proportion of steam relative to liquid-phase water in the core would produce an increase in reactivity. The reactor became super-critical.

This positive coefficient of reactivity was a design weakness of the RBMK reactor. In contrast, light water reactors of the BWR type have a *negative* void coefficient, that is, the reactivity declines when there is an above-normal amount of boiling in the core. This is because the water serves as both coolant and moderator, so that when voiding occurs the effect is to reduce the amount of moderation, and hence the rate at which neutrons are absorbed in fissile nuclei. But in the RBMK reactor moderation is provided by the graphite and so is unaffected by voiding in the water coolant. Under the conditions of the test that was carried out at Chernobyl, the overriding effect of the steam voiding was to increase the absorption of neutrons in the fissile ^{235}U and hence to increase the reactivity.

When the Chernobyl reactor went super-critical, the power started to rise, causing a further increase in voiding, followed by a further increase in reactivity. A nuclear runaway ensued, and the power level began to increase very rapidly. By one estimate, it had reached 100 times the normal full power level within four seconds. There was not enough time to stop the chain reaction by moving the control rods down into the core. The passage of the control rods was in any case blocked by the destruction of the fuel channels caused by the huge power surge. The fuel started to overheat, melt, and fragment. As the steam came into contact with the pieces of hot fuel there was a steam explosion that was powerful enough to lift off the 1,000-ton upper plate of the reactor. A second explosion occurred shortly afterwards, which was caused by hydrogen produced in chemical reactions between the steam and the fuel cladding and the red-hot graphite. This second explosion blew large amounts of radioactivity into the atmosphere. The hot graphite that was still in the core was exposed to air and ignited, and graphite fires in and around the core continued to burn for days. Further melting occurred in the destroyed core as the fission product decay heat built up, and large quantities of radionuclides continued to be released into the atmosphere. The situation was not stabilized for many days, and even then only after heroic (and for some of those involved, ultimately fatal) interventions by firefighters and other emergency workers.

Investigations of the accident and its aftermath apportioned blame in various combinations to administrative negligence and incompetence, exacerbated by ingrained secrecy, operator error, and basic design flaws in the RBMK reactor. The disaster at Chernobyl did not mean the end of nuclear power in the countries of the former Soviet Union, but development was slowed greatly. RBMK reactors have continued to operate, but no new reactors of this type have been built since the accident, and the Russian nuclear program is now based on PWRs.

The global impact of Chernobyl on nuclear energy was great. It affected public attitudes towards nuclear safety around the world more profoundly than any other single event either before or since. And even as officials in the West, Japan, and other countries sought to downplay the relevance of Chernobyl to the safety of their own nuclear power programs, they were well aware of the extreme vulnerability of these programs to another accident of comparable magnitude, wherever it occurred. International cooperation to promote nuclear safety intensified in the aftermath of Chernobyl, but in the court of public opinion the nuclear energy option had sustained serious damage.

In the next two chapters, we turn to two other longstanding problems that are closely identified with nuclear power: the disposal of nuclear waste and the proliferation of nuclear weapons. As we shall see, both issues will require a great deal of careful attention irrespective of whether the nuclear energy option gains a new lease on life. On the other hand, a failure to manage both of them effectively will very likely eliminate any chance of a nuclear revival. To prepare for this discussion, the final section of the current chapter introduces the key stages of the nuclear fuel cycle.

The Nuclear Fuel Cycle

A complex cycle of industrial operations is required to prepare and manufacture fresh fuel for nuclear power reactors and to manage "spent" (irradiated) fuel after it is discharged. The particular characteristics of the nuclear fuel cycle depend on the type of reactor that is being supported. Here we concentrate on the fuel cycle for light water reactors (LWRs), which, as noted previously, use uranium fuel that is enriched to about 3% in the fissile isotope U-235. A schematic diagram of the LWR "once-through" fuel cycle is shown in Figure 7.3. (In the once-through cycle, the spent fuel is disposed of directly with no effort to recover unused fissile isotopes.)

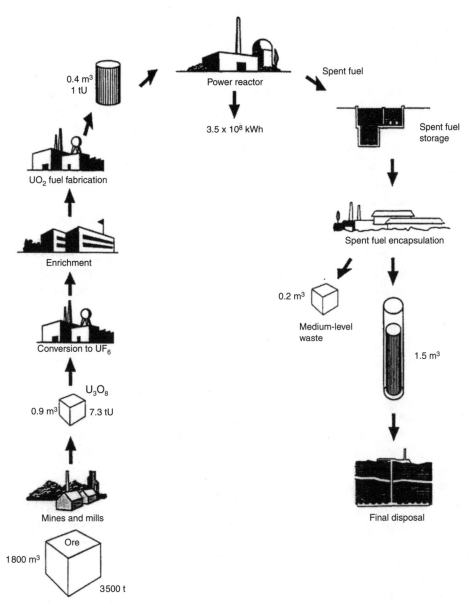

0.4 m³
1 tU

Power reactor

Spent fuel

3.5 x 10⁸ kWh

Spent fuel storage

UO₂ fuel fabrication

Spent fuel encapsulation

Enrichment

0.2 m³

Medium-level waste

1.5 m³

Conversion to UF₆

U₃O₈

0.9 m³   7.3 tU

Mines and mills

Final disposal

Ore

1800 m³

3500 t

Source: Illustration by the Nuclear Energy Agency, *The Economics of the Nuclear Fuel Cycle*, OECD, Paris, France, 1994, Table 3.2.

Figure 7.3. Schematic of once-through LWR fuel cycle, with approximate volumetric and mass flows (basis: 1 metric ton of low-enriched uranium fuel).

The Front End of the Fuel Cycle

Uranium mining is the first stage of the fuel cycle. Uranium ore deposits occur in many parts of the world. The main producing nations today are the United States, Australia, South Africa, Canada, Russia, and other nations of the former Soviet Union.

Uranium ore typically contains only a few tenths of a percent of uranium, and the ore is processed in a uranium mill to produce "yellowcake," a concentrate containing 85% to 90% by weight of uranium oxide (U_3O_8). The mill is usually located close to the mine site in order to minimize the cost of transporting the ore. The nonuraniferous material, which constitutes the vast bulk of the ore, is rejected at the mill. This material, known as the mill tailings, contains most of the radioactive daughter products of uranium that were present in the ore, and must be stabilized to prevent the release of these radioisotopes (including radon gas) into the environment.[4]

In the next stage of the cycle, the yellowcake is purified and converted to uranium hexafluoride (UF_6), the only chemically stable compound of uranium that is volatile at temperatures close to ambient. UF_6 is the feed material for the next stage, uranium enrichment, in which the weight fraction of the fissile isotope ^{235}U is increased from 0.711% up to about 3% – the fissile concentration needed for LWR fuel. The isotopic enrichment of uranium is one of the most technically challenging stages of the fuel cycle. Straightforward chemical separation is impossible, because different isotopes of the same element exhibit identical chemical behavior, and several ingenious (and expensive) physically based separation schemes have been developed over the years to accomplish this task.

The two main enrichment technologies in commercial use today are gaseous diffusion and the gas centrifuge process. For several decades, gaseous diffusion plants produced almost all of the enriched uranium used in nuclear power reactors, and still today account for most of the world's enrichment capacity. The process relies on the slight (less than 1%) mass difference between molecules of $^{235}UF_6$ and $^{238}UF_6$. Gaseous UF_6 is pumped under pressure across a semi-porous diffusion barrier. The lighter $^{235}UF_6$ molecules have a slightly higher probability of diffusing through the barrier, and the gas on the downstream side is thus slightly enriched in the fissile isotope, while the undiffused gas is slightly depleted (see Figure 7.4).

The ratio of U-235 to U-238 in the downstream gas is only slightly higher than in the feed gas, and more than 1,000 stages are needed to achieve a U-235 enrichment of 3%. The performance of each enrichment stage is described by the separation factor, α, given by the expression:

$$\alpha = \frac{\frac{x_P}{1-x_P}}{\frac{x_W}{1-x_W}}$$

where x_P and x_W are the weight fractions of U-235 in the enriched and depleted

---

[4] Recall that both of the naturally occurring isotopes of uranium – ^{235}U and ^{238}U – are radioactive and undergo long half-life α-decay reactions to produce daughter products that are themselves radioactive, as are their daughters. In each case, in fact, the initial decay is followed by several successive (mostly alpha) decay events until finally a stable, nonradioactive isotope results. In natural uranium ore, the radioactive intermediate daughter products in these decay chains exist alongside (and in radioactive decay equilibrium with) their "parents," the ^{235}U and ^{238}U isotopes. It is these radioactive daughter products that show up in the tailings after the uranium is removed in the mill.

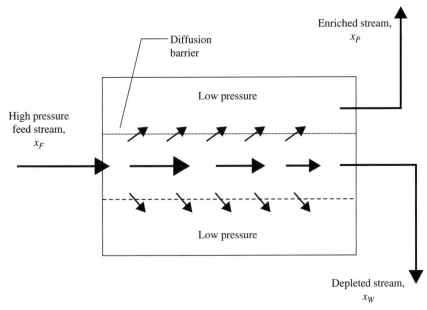

Figure 7.4. Schematic of a gaseous diffusion enrichment stage.

product streams, respectively. The stage separation factor for a gaseous diffusion stage is 1.00429. (An analogous separation factor is used to characterize other isotope separation processes, too.)

The stages are arranged in a "cascade," in which the enriched product from one stage becomes the feed to the next highest stage, while the depleted product becomes the feed to the next lowest one. The feed stream is introduced into a central stage of the cascade, while the enriched product and depleted "tails" streams are withdrawn from each end (see Figure 7.5).

An overall material balance on the cascade yields:

$$F = P + W, \tag{4}$$

and a material balance on the U-235 isotope leads to

$$F x_F = P x_P + W x_W, \tag{5}$$

where F, P, and W are the masses of uranium in the feed, product, and tails streams respectively, and x_F, x_P, and x_W are the weight fractions of U-235 in the three streams. For a cascade enriching natural uranium to 3% in U-235 at a tails assay, x_W, of 0.2%, solving equations (4) and (5) gives:

$$F = P \left(\frac{x_P - x_W}{x_F - x_W} \right) = 5.48 P,$$

that is, to produce 1 kilogram of 3% enriched uranium product requires about 5.5 kilograms of natural uranium feed if the cascade is operating with a tails assay of 0.2%. The feed requirement per unit of product increases as the tails assay is raised.

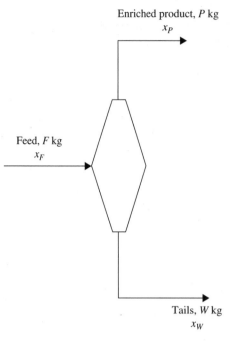

Enriched product, P kg
x_P

Feed, F kg
x_F

Tails, W kg
x_W

Figure 7.5. Schematic of an enrichment cascade.

Gaseous diffusion plants are extremely large, very capital intensive, and use large amounts of energy. A full-scale gaseous diffusion plant consumes 2,000–3,000 megawatts of electric power, enough to meet the needs of a city of million or more people. Commercial-scale plants are today operating in the United States, Russia, and France.

The gas centrifuge process, the other leading enrichment technology, also relies on the small mass difference between molecules of $^{235}UF_6$ and $^{238}UF_6$. In this case, the separation is achieved in ultra-high-speed centrifuges. UF_6 gas introduced into the centrifuges is subject to centrifugal acceleration thousands of times greater than gravity. The heavier $^{238}UF_6$ molecules tend to congregate at the centrifuge wall, while the gas at the axis is enriched in $^{235}UF_6$. Two separate gas streams are removed from the centrifuge, one from the central region (slightly enriched) and one from the edge region (depleted.) The overall separation factor, α, in an optimally designed centrifuge is roughly 1.4 – much higher than in a gaseous diffusion stage. However, the throughput of each centrifuge is small, because of materials and mechanical constraints that limit the size and rotation speed of the machine. To produce commercial-scale quantities of enriched uranium, hundreds or thousands of centrifuges must be piped together in a cascade. Gas centrifuge cascades are even more capital-intensive than gaseous diffusion plants, but only consume about 5% of the energy. An Anglo-Dutch-German consortium operates the only full-scale gas centrifuge enrichment complex in service today.

Several other enrichment technologies have been developed to pilot scale over the years, but none has been deployed commercially. In each case the technical and economic hurdles have turned out to be greater than anticipated. The uncertain future of nuclear power has also discouraged investment in new enrichment technologies in recent years. In the United States, the leading alternative technology, the atomic vapor laser isotope separation process (AVLIS), was recently dropped after a twenty-five year development effort. Development work on various enrichment processes continues in several countries around the world, however, and because of the potential of these technologies to produce highly enriched uranium suitable for use in nuclear weapons some of these activities raise serious security concerns. For example, the nuclear explosions staged by Pakistan in the 1990s used highly enriched uranium produced in a small-scale gas centrifuge plant. We discuss the problem of managing the nuclear weapons proliferation risks associated with nuclear power technology in more detail in Chapter 9.

The next stage in the nuclear power fuel cycle involves the fabrication of nuclear fuel assemblies. The 3%-enriched UF_6 is first converted to uranium oxide, UO_2. The uranium oxide powder is then formed into small cylindrical pellets about half an inch long and half an inch in diameter. The pellets are stacked one on top of the other inside thin-walled zircaloy tubes 12 ft in length. The zircaloy cladding protects the uranium from corrosion by the hot cooling water inside the reactor and also helps to prevent the release of highly radioactive fission products into the coolant. A typical PWR fuel assembly consists of 289 of these fuel "pins" bundled together in a square 17 × 17 array. The core of the reactor consists of a large number of these fuel assemblies. The fuel typically stays in the reactor for three years; a third of the assemblies are removed and replaced during the annual refueling shutdown.

A 1,000 MWe light water reactor, operating with an annual capacity factor of 80% and a thermal efficiency of 33%, produces

1000 (MWe) × 1/0.33 (MWth/MWe) × 365 (days/yr) × 0.8 = $8.76 × 10^5$ MWDth/yr.

Since the fissioning of 1 g of U-235 yields roughly 1 megawatt-day of thermal energy, this implies that the reactor needs 876 kilograms of U-235 each year. If the U-235 enrichment of the fresh fuel is 3%, this would be equivalent to an annual fuel requirement of 876/0.03 = 29,200 kg of enriched uranium per year, or 29.2 MTHM (metric tonnes, heavy metal).

This calculation is not exact, since not all of the U-235 is fissioned before the fuel is discharged. (Typically, the U-235 enrichment in spent PWR fuel is about 0.9%.) On the other hand, as irradiation continues, an increasingly important contribution to overall energy production is made by the fissioning of plutonium-239, which is formed by the absorption of neutrons in U-238.

A useful measure of the in-core performance of nuclear fuel is its cumulative production of thermal energy, or "burnup," at the time of discharge. For 3% enriched light water reactor fuel, a typical design burnup is 33,000 megawatt-days per metric ton of heavy metal (MWDth/MTHM). Assuming a discharge U-235 enrichment of 0.9%, the energy released from the fissioning of the U-235 in the fuel

$$= (0.03 - 0.009) × 10^6 \text{ (g U-235/MTHM)} × 0.95(\text{MWD/g})$$
$$= {\sim}20,000 \text{ MWD/MTHM},$$

that is, a little under two-thirds of the total. Most of the remainder is accounted for by Pu-239 fissions. (By the time the fuel is ready to be discharged, Pu-239 fissions are responsible for about half the total power output.)

A material balance on the front-end of the nuclear fuel cycle for a 1,000 MWe PWR is presented in Figure 7.6. Assuming a spent fuel burnup of 33,000 MWD/MTHM, a 1,000 MWe reactor operating at an annual capacity factor of 80% and a thermal efficiency of 33% requires

$$1{,}000 \ (\text{MWe}) \times 365 \ (\text{days/yr}) \times 0.8 \times (1/0.33)(\text{MWth/MWe})$$
$$\times (1/33{,}000)(\text{MTHM/MWDth}) = 26.55 \ \text{MTHM/yr of fresh fuel.}$$

Working back from the reactor, the material balance shows that about 175 metric tons of yellowcake are required each year to meet the reactor's annual fuel requirement. (It is interesting to compare this with the requirement of a 1,000 MWe coal-fired power plant for about 10,000 metric tons of coal each day.)

The Back End of the Fuel Cycle

After discharge from the reactor, the spent fuel is initially stored in a water-filled pool adjacent to the reactor vessel. The fuel is hot and highly radioactive, and the water both cools the fuel and provides radiation shielding protection. Most spent fuel at U.S. power reactors remains in these storage pools. At some reactors, however, the pools have filled up, and to make room for newly discharged fuel batches some of the older assemblies (which generate less heat as they age) have been placed in air-cooled concrete storage casks. Eventually the spent fuel will be moved offsite, either to a centralized temporary storage facility or directly to a geologic repository for final disposal. These options are discussed in more detail in the next chapter.

The utilities in the United States and several other countries have adopted the once-through fuel cycle in which the spent fuel will be disposed of directly. A few other countries have opted to reprocess their spent fuel in order to recover the Pu-239 that it contains, as well as the uranium (which, recall, is still slightly enriched in U-235). The principal reprocessing plants now operating are in France and the United Kingdom, and utilities in Europe and Japan are sending some of their spent fuel to these facilities.

The primary technology for spent fuel reprocessing is the PUREX process. The spent fuel rods are chopped into short sections and then dissolved in hot nitric acid, and the plutonium and uranium in solution are recovered and purified through a succession of solvent extraction stages. The primary waste stream from reprocessing is an aqueous solution containing almost all of the fission products, trace quantities of uranium and plutonium not recovered during solvent extraction, and relatively small quantities of several other actinide species (mainly neptunium, americium, and curium) formed by neutron absorption reactions in uranium and plutonium. This highly radioactive waste stream is stored temporarily in stainless steel tanks to allow further decay and is then converted to water-insoluble borosilicate glass logs, stored in cylindrical canisters.

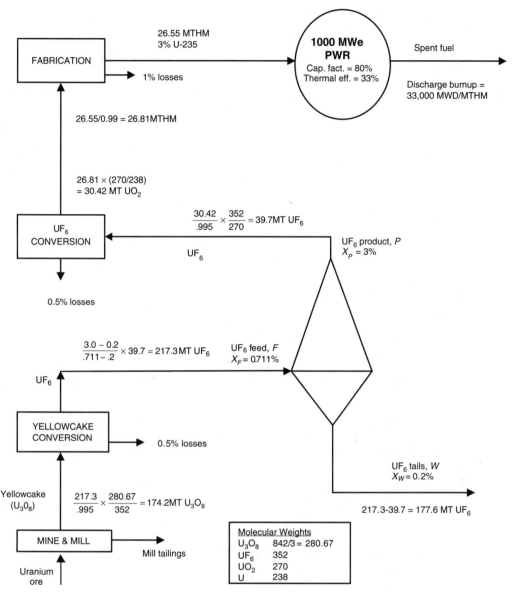

Figure 7.6. Material balance on the front end of the PWR fuel cycle (basis: one year of operation).

Eventually the high-level waste canisters will be shipped to a permanent repository. A simplified flow diagram of the PUREX process is shown in Figure 7.7.

In a proposed variant of this scheme, the higher actinides would be partitioned from the high-level waste stream prior to solidification. Some of the actinides have half-lives of tens of thousands of years or more – much longer than the half-lives of most of the fission products. Separating them from the rest of the waste could thus significantly reduce the time for which the latter would have to be isolated from the

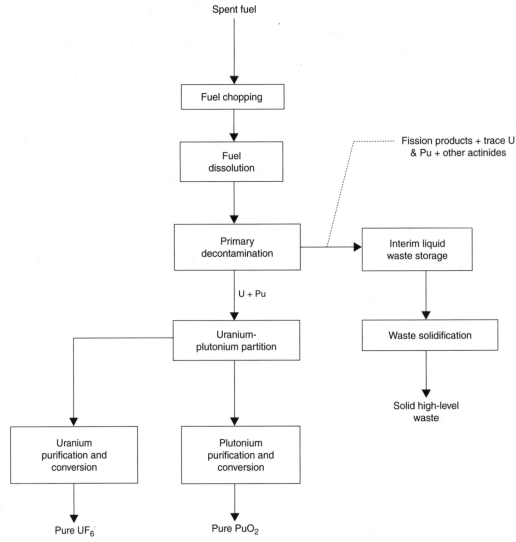

Figure 7.7. Simplified PUREX process flow diagram.

biosphere. The separated actinides can in principle be converted to shorter-lived fission products by exposing them to a neutron flux in a fission reactor or, alternatively, in an accelerator-driven device. (This process, known as actinide transmutation, is feasible because several of the actinide species are fissionable.)

Most of the operating and maintenance activities in reprocessing plants must be conducted remotely because of the intense radiation fields involved. The key steps are carried out in heavily shielded process cells, enclosed within massive reinforced-concrete walls or "canyons." To reduce the risk of misuse of separated plutonium, current proposals call for modifications to the standard PUREX flowsheet that would preclude the complete separation of plutonium from uranium and instead produce a uranium-plutonium mixture with the plutonium concentration in the blend not exceeding the level needed for subsequent fuel recycle.

At present some of the separated plutonium from the British and French reprocessing plants is being recycled to LWRs in mixed-oxide (MOX) fuel – uranium oxide fuel containing 2% to 5% by weight of plutonium oxide. Because of the special radiation and criticality problems associated with handling plutonium, MOX fuel assemblies must be fabricated in dedicated facilities that are significantly more costly to build and operate than standard LWR fuel fabrication plants. The high cost of reprocessing the spent fuel to extract the plutonium and the additional costs of fuel fabrication make MOX fuel considerably more expensive than "conventional" low-enriched uranium fuel. Despite this, some advocate the use of MOX fuel in LWRs as a way to work off the growing stock of separated plutonium, which must be very carefully protected to guard against its use in nuclear weapons. An alternative use for the separated plutonium is to provide the start-up cores for fast breeder reactors, but at present the economic case for breeder reactors is even weaker, and there are no plans anywhere in the world to proceed with the commercial deployment of this technology.

Large amounts of separated plutonium from reprocessing spent LWR fuel are accumulating around the world, and the effective control of this material, and of the reprocessing technology that can produce more of it, is a policy goal of great importance. We will discuss the problem of nuclear weapons proliferation and its relationship to nuclear power in more detail in Chapter 9.

8

Managing Nuclear Waste

The management and disposal of radioactive waste from the nuclear fuel cycle is one of the most intractable problems facing the nuclear power industry around the world. Today, more than forty years after the first nuclear power plant entered service, no country has yet succeeded in disposing of high-level nuclear waste – the longest-lived, most highly radioactive, and most technologically challenging of the waste streams generated by the nuclear industry. Most countries have stated their intention to dispose of the waste in repositories constructed in rock formations hundreds of meters below the earth's surface. But no country has actually put a geological repository for high-level waste into service, and all have encountered difficulties with their programs. The problems are partly technological and partly political, and it is impossible to draw a sharp line between them.

The basic policy questions in the field of high-level nuclear waste – What is to be done? Who should decide? Who should pay? Who will implement the solution? – are questions encountered in many other fields of technology. In this case, however, their resolution is made more difficult by several factors, including the very hazardous nature of the waste itself, the extremely long time for which it must be contained, and the special fear that it evokes among many people as a result of these characteristics and the disturbing presence of nuclear radiation. Another complicating factor is that the waste is physically quite compact, so that a single repository will be large enough to store the waste from many power reactors. Though obviously advantageous in one sense, this also creates an inherent "spatial" inequity in any solution – those living near proposed repository sites perceive themselves to be, and in fact are, carrying the burden for a much larger population. Third, the management and disposal of waste is part of a larger system, the nuclear power fuel cycle, and what is decided here cannot be decoupled from issues and decisions elsewhere in the cycle concerning, for example, nuclear nonproliferation or the abatement of greenhouse gas emissions. This interconnectedness adds to the complexity of decision-making in the waste management field. Because of these difficulties, almost all the leading nuclear countries expect that it will take decades to establish operating high-level waste repositories (see Table 8.1).

Our purpose in this chapter is to examine the role of the technologist in this complicated domain. How can the practicing scientist or engineer contribute most effectively in a field with so many challenging nontechnological as well as technological issues? Does good technical practice mean leaving the nontechnical issues for others to deal with? If not, how should the technologist address them? As we will see, good technical practice requires the technologist to take a holistic view of the problem, to understand the often

Table 8.1. High-level waste disposal plans of leading nuclear countries

Country	Preferred geologic medium	Earliest anticipated repository opening date	Status
United States	Volcanic tuff	2010	Site selected (Yucca Mountain, NV); application for construction license in preparation
Finland	Crystalline bedrock	2020	Site selected (Olkiluoto, SW Finland) – decision ratified by Parliament in May 2001
Sweden	Crystalline rock	2020	Searching for a suitable site
Switzerland	Crystalline rock or clay	2020 or later	Searching for a suitable site
France	Granite or clay	2020 or later	Developing repository concept
Canada	Granite	2025 or later	Reviewing repository concept
Japan	Not selected	2030	Searching for suitable site
United Kingdom	Not selected	After 2040	Delaying decision until 2040
Germany	Salt	No date specified	Moratorium on repository development for 3–10 years

very different perspectives that others bring to bear, and to make careful assessments of where the application of scientific and engineering methods can help provide solutions, and where they cannot. All of this makes the technologist's task much more difficult. But the alternative approach, of ignoring those issues that are not strictly technical in nature, will very likely lead to technological failure.

THE TECHNICAL CHARACTERISTICS OF HIGH-LEVEL WASTE

Our focus in this chapter is on high-level waste. This takes the form either of spent fuel assemblies or, if the fuel is reprocessed, the primary waste stream from the reprocessing plants. The latter contains almost all of the fission products and most of the actinides in the spent fuel except plutonium and uranium, which are recovered separately.

In the United States, as discussed in the previous chapter, there are no current plans to reprocess the spent fuel from nuclear power reactors, and most of this material is now stored in water-filled basins at the reactor sites. In the previous chapter we saw that a typical 1,000 MWe light water reactor discharges 25–30 MTHM (metric tons of heavy metal) of spent fuel per year. These assemblies occupy a volume of about 10 m^3. If we assume an average reactor operating lifetime of thirty years, the total lifetime spent fuel inventory from the U.S. fleet of about 100 power reactors will be 75,000–90,000 MTHM, which would occupy a volume equivalent to a cube measuring roughly 100 ft along a side. (Of course, in practice the fuel assemblies could not be packed so closely together; each assembly must be cooled adequately to prevent the radioactive decay heat from raising the fuel temperature to unsafe levels.) Today about

40,000 MTHM of spent fuel is in storage at the reactor sites with about 2,000 MTHM being added every year.

The other source of high-level waste in the United States is the atomic energy defense activities of the federal government. Since World War II, government nuclear reactors producing plutonium and tritium for the nuclear weapons program have discharged large amounts of spent fuel, most of which has been reprocessed. Large amounts of spent fuel from the naval nuclear reactors program have also been reprocessed. The liquid high-level waste from these reprocessing plants has been stored in tanks at the Hanford Reservation in eastern Washington, at the Idaho National Laboratory, and at the Savannah River Site in South Carolina.

The concentration of radionuclides in the government's reprocessed high-level waste is lower than in spent fuel, but the volume is much larger – more than 300,000 m^3, compared with a current commercial spent fuel inventory of less than 20,000 m^3. Storing this material in tanks is at best a temporary solution. At Hanford about a third of the 177 high-level waste tanks have leaked, and local groundwater has been contaminated. Leaks have also occurred at Savannah River. Most of the waste in these tanks has now been solidified. At all three sites, the waste will have to be removed, processed, and immobilized in a form acceptable for permanent disposal. This will be a very costly task. At Hanford alone the cost of cleaning up the tanks is expected to exceed $30 billion. The cost estimates for the Savannah River Site are comparable. (The high-level waste at the Idaho site is in a different physical and chemical form and will be more straightforward to deal with.) The government plans eventually to dispose of the high-level waste from its own sites in the same repository that will contain the commercial spent fuel.

Parenthetically, it is important to note that several other types of radioactive waste are also generated in the nuclear fuel cycle (in larger volumes than high-level waste), as well as by many other industrial, medical, and research activities involving the handling of nuclear materials. These wastes vary widely in physical and chemical form, handling requirements, and the length of time for which they must be isolated from the biosphere. Transuranic wastes are defined as non-high-level wastes contaminated with small quantities of alpha-emitting transuranic isotopes with half-lives greater than twenty years. These wastes are produced primarily in fuel reprocessing and mixed-oxide fuel fabrication. Mill tailings are the residues from uranium ore mining and milling operations that contain low concentrations of the naturally-occurring radioactive daughter products of uranium. Low-level wastes are defined by exclusion to consist of everything that is not high-level or transuranic wastes – a category that covers a wide range of physical and chemical forms and radiation levels. Although the high-level wastes that are the focus of this chapter pose the most difficult technical challenges, managing the other types of nuclear waste has also proved to be very demanding task in most countries. Current U.S. inventories of each category of nuclear waste are shown in Figure 8.1.

The most important physical characteristics of high-level waste are the radioactivity and radioactive decay heat that it emits. These two properties strongly influence the shielding, cooling, and containment requirements at each stage of the waste management system, including temporary storage, transportation, and the repository

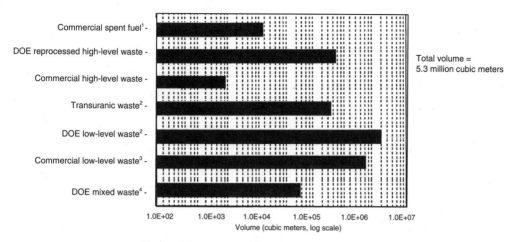

Commercial spent fuel[1]

DOE reprocessed high-level waste

Commercial high-level waste

Transuranic waste[2]

DOE low-level waste[2]

Commercial low-level waste[3]

DOE mixed waste[4]

Total volume =
5.3 million cubic meters

1.0E+02   1.0E+03   1.0E+04   1.0E+05   1.0E+06   1.0E+07
Volume (cubic meters, log scale)

[1] Permanently discharged reactor fuel includes spacing between fuel assembly rods.
[2] DOE wastes include both retrievably stored and buried materials.
[3] Includes contributions from disposed wastes only.
[4] Includes retrievably stored RCRA and TSCA materials only.

Source: U.S. Department of Energy, "Integrated Database Report – 1998: U.S. Spent Nuclear Fuel and Radioactive Waste Inventories, Projections, and Characteristics", DOE/RW-0006

Figure 8.1. U.S. nuclear waste inventories – 1998.

itself. The decay behavior of spent fuel from a pressurized water reactor is shown in Figure 8.2. Note the logarithmic scale of these graphs. The left-hand panel shows that the total radioactivity declines by a factor of about 1,000 between one and 1,000 years and continues to decline rapidly thereafter. During the period beginning a few years after discharge from the reactor and spanning the first few hundred years, the dominant contributors are two fission products – strontium-90 (twenty-eight-year half-life) and cesium-137 (thirty-year half-life). After about 300 years, or a span of roughly ten half-lives, the radioactivity of these isotopes has decayed by a factor of $(1/2)^{10} \approx (1/1000)$, and by then the transuranic isotopes have taken over as the dominant contributors to the radioactivity. The thermal decay behavior in the right-hand panel of Figure 8.2 follows a similar trajectory. During the first several decades, the fission products are again the primary contributors, with the long half-lived transuranics later becoming dominant.

Over these long time-scales the rate of heat release is modest. For example, a typical PWR fuel assembly, which produces about 20 megawatts of fission power while in the reactor core, generates only about 2 kw of decay heat ten years after shutdown. After 100 years, the decay heat declines to about 500 watts. But even these low thermal power levels can cause significant temperature increases in underground waste repositories and are an important consideration in designing these facilities. The elevated temperatures and thermal gradients may affect the movement of groundwater in the vicinity of the repository, for example, by creating fractures in the rock that might then admit groundwater or by inducing convective flow. The rate of corrosion of waste canisters is likely to be accelerated at higher temperatures too.

Another way to characterize the waste is in terms of the risk it poses to humans. This declines over time as the radionuclides decay. One measure of this risk is the so-called

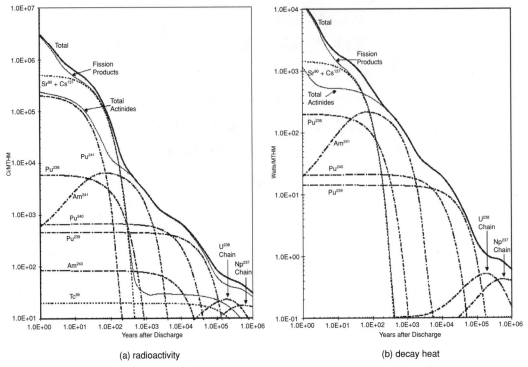

Figure 8.2. Decay behavior of spent PWR fuel (basis: 1 metric ton of fuel; initial enrichment of 4.5% ^{235}U; discharge burnup of 50 MWD per MT of heavy metal.)

"ingestion hazard index," defined as follows. For each radioisotope, the radiation protection authorities have specified a maximum allowable concentration of the isotope in water such that an individual could safely obtain his total water intake from such a source. The ingestion hazard index is then the total volume of water required to dilute all of the radionuclides in a unit mass of waste down to their maximum permissible concentrations. Thus, for 1 MT of waste:

$$\text{Ingestion hazard index at time } t \; (\text{m}^3/\text{MT}) = \sum_i^{all \; radionuclides} \left(\frac{\lambda_i N_i(t)}{MPC_i^{water}} \right),$$

where $\lambda_i N_i(t)$ is the amount of radioactivity of isotope i present in one ton of waste at time t (in Bq/MT), and MPC_i is the maximum permissible concentration of isotope i in water (in Bq/m^3.)

Figure 8.3 shows how the ingestion hazard index of one metric ton of spent PWR fuel declines over time. Also shown for comparison is the index for the equivalent amount of natural uranium ore – that is, the quantity of uranium ore that would have to be mined in order to generate the metric ton of spent fuel. According to the figure, after about 10,000 years the spent fuel will be no more hazardous than the parent ore, implying that a high-level waste repository should be designed to isolate the spent fuel for approximately that length of time. Of course, such comparisons take no account of the different environmental risk factors for these materials. Uranium ores (and other naturally occurring hazardous materials) are deposited randomly, frequently in permeable strata,

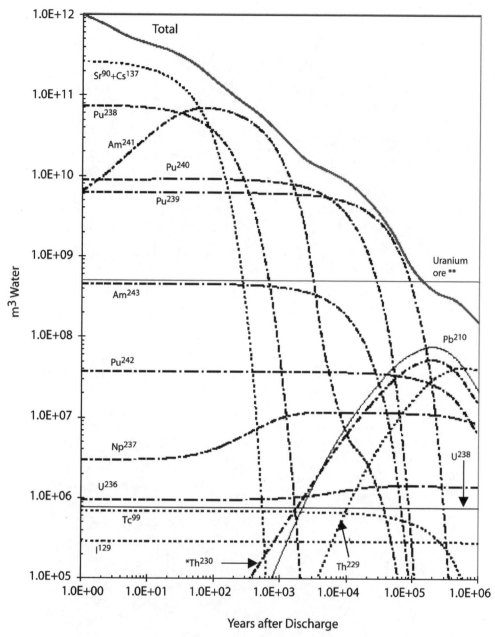

Figure 8.3. Ingestion hazard index of spent PWR fuel (basis: 1 metric ton of fuel; initial enrichment of 4.5% ^{235}U; discharge burnup of 50 MWD per MT of heavy metal.)

and with groundwater often present in abundance. By contrast, high-level waste will be buried at depths of several hundred meters in locations selected for geological stability, low groundwater flows, and remoteness from population centers. On the other hand, a high-level waste repository is a manmade structure, with shafts and boreholes linking

Source: Courtesy U.S. Department of Energy, Office of Civilian Radioactive Waste
Management, Yucca Mountain Project website, http://www.ymp.gov/.

Figure 8.4. Aerial view of the crest of Yucca Mountain, Nevada.

it to the biosphere. Moreover, as noted previously the presence of heat-generating
materials has the potential to disrupt the geohydrological environment and accelerate
the corrosion of the waste canisters. All of these factors – and others besides – must be
considered in assessing the actual risk posed by a waste repository.

Source: Courtesy U.S. Department of Energy, Office of Civilian Radioactive Waste Management,
Yucca Mountain Project website, http://www.ymp.gov/.

Figure 8.5. Artist's rendition of the Yucca Mountain repository.

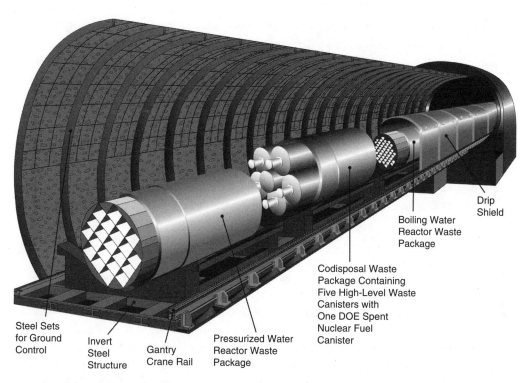

Steel Sets for Ground Control

Invert Steel Structure

Gantry Crane Rail

Pressurized Water Reactor Waste Package

Codisposal Waste Package Containing Five High-Level Waste Canisters with One DOE Spent Nuclear Fuel Canister

Boiling Water Reactor Waste Package

Drip Shield

Source: Illustration by U.S. Department of Energy, Office of Civilian Radioactive Waste Management, Yucca Mountain Project website, http://www.ymp.gov/.

Figure 8.6. Cutaway of emplacement tunnel proposed for Yucca Mountain repository containing three types of waste packages.

Figure 8.4 shows the site at Yucca Mountain in Nevada that has been selected as the location for the first U.S. high-level waste repository (the selection process is discussed in more detail later in the chapter). The host rock at Yucca Mountain is tuff, a rock formed from volcanic ash, and is estimated to be about 12 million years old. An artist's rendition of the repository is shown in Figure 8.5. In the proposed design, nuclear fuel assemblies will be contained in metal alloy canisters, which will in turn be encapsulated in outer canisters made of steel. The waste containers will be transported into the repository on rail carts down gently sloping access tunnels bored into the side of the mountain. The containers will be stored horizontally in tunnels some 300 meters below the surface. The conceptual design of one of these tunnels, or "drifts," is shown in Figure 8.6. After a fifty-year period of monitoring to ensure that the containers are behaving as planned, each emplacement tunnel will be backfilled and closed. Eventually, the entire repository will be backfilled and sealed off from the biosphere.

A repository must provide protection against every plausible scenario in which radionuclides might reach the biosphere and expose the human population to dangerous doses of radiation. Various possibilities must be considered, including the risk of volcanic activity and the possibility of human intrusion into the repository, either inadvertent or intentional. Of all possible pathways, the one receiving most attention involves the flow of groundwater into the repository, the corrosion of the waste containers, the leaching of radionuclides into the groundwater, and the migration of the contaminated groundwater towards locations where it might be used as drinking water or for agricultural purposes.

The repository must be designed to minimize the likelihood of such a scenario. The waste containers must be made of highly corrosion-resistant materials. The repository itself should be located in a region where the rate of groundwater infiltration will be low, and where the groundwater travel path from the repository to actual or potential potable water supplies will be long. Ideally, the rock surrounding the repository should also be of a type to retard the migration of radionuclides by absorbing them on the rock surfaces as the groundwater seeps through. The repository should be located in a geologically stable region so as to minimize the risk of new groundwater pathways opening up as a result of tectonic activity, and the site should be distant from human population centers.

An important design consideration concerns the relative contributions of the different barriers to radionuclide migration in the repository system. In some approaches, primary reliance is placed on the natural geologic barriers. The waste canister and other engineered components are expected to play an important supporting role, but the main burden of containment is shouldered by the repository host rock formation and the surrounding geohydrological region. In other concepts, more emphasis is placed on developing long-lived engineered barriers. In the Swedish program, for example, the primary containment barrier throughout the repository lifetime will be thick-walled copper and steel canisters, with the host rock serving as a backup in the event that the canisters should fail. The difficulty with relying on engineered structures is that there is very little experience with the performance of any human artifacts for periods longer than a few hundred years. In the case of geologic barriers the historical record is obviously much longer, often extending over millions of years, but such natural structures cannot be characterized with anything like the precision of engineered structures, and the historical record is subject to considerable interpretation. The challenge for repository designers will be to develop overall containment systems that combine the strengths of each kind of barrier in a way that effectively compensates for their weaknesses.

CONTRASTING PERSPECTIVES ON THE NUCLEAR WASTE MANAGEMENT PROBLEM

For reasons to be discussed in more detail below, the United States may not have an operating high-level waste repository for another twenty years or more, and spent fuel continues to accumulate at the seventy or so operating nuclear power plant sites around the country. When these plants were built it was expected that the spent fuel would be shipped off to reprocessing plants within a year or so of being discharged from the core, and so the reactor storage pools were built with limited capacity. These pools are rapidly filling up, and in some cases the capacity has already been exhausted (see Figure 8.7).

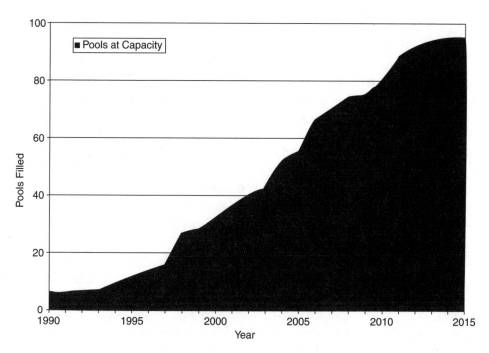

Source: Reprinted from M. S. Kazimi and N. E. Todreas, "Nuclear Power Economic Performance: Challenges and Opportunities," *Annual Reviews of Energy and the Environment*, 24 (1999), 139–171.

Figure 8.7. Projected exhaustion of U.S. reactor fuel pool storage capacity.

At some of these sites some of the older, cooler fuel assemblies have been transferred to air-cooled, steel storage casks sitting on concrete pads.

Although these "dry" casks are expected to have a lifetime of several decades, the solution is only a temporary one. Moreover, at some locations there is not enough room for storage casks. If there is nowhere to store the spent fuel, some reactors may be forced to shut down prematurely.

How would an engineer, using the engineer's arsenal of problem-solving techniques, deal with these bottlenecks at the back-end of the nuclear fuel cycle? We might expect that he or she would approach the back-end as a *system* – a network of fuel pools, casks, processing facilities, and disposal sites connected by road and rail transportation links – and try to optimize the configuration and design of this system with respect to both safety and cost. The engineer would evaluate the short-term risks of storing the fuel in pools or dry storage casks, and almost certainly find that there was no urgent need to dispose of the waste on safety grounds. She or he might then conclude that the best strategy would be to search systematically for the best possible disposal site or sites, while investing prudently in the development of new information concerning the behavior of waste in geologic repositories to ensure that the eventual design of the repository would be based on the best possible knowledge base. Because all of this would likely take time, a strategy for temporary storage of the spent fuel would be required in the meantime. To avoid the premature shutdown of some reactors, to exploit economies of scale in storage operations, and to provide better protection against the threat of sabotage, this would probably entail developing one or more central storage facilities. An important consideration in siting these facilities would be to minimize transportation costs and risks.

In the real world, of course, the engineer does not practice his profession in isolation. Multiple interests are at stake, multiple parties are involved in decision-making, and the different parties approach the problem with different assumptions, different goals, and different ways of thinking. Environmentalists, for example, do not see the waste management program as a problem of system optimization. For the environmental activist, a waste repository is a blight and not something to be optimized. Even the best-designed repository will be an insult to the environment. The repository design engineer, seeking to minimize the risks to both current and future generations, prefers a location as remote as possible from population centers. But the ardent environmentalist, dedicated to the preservation of wilderness areas, focuses on the disruptions that such a facility will cause – the large-scale subsurface construction activities, the massive surface facilities, and the continuing operations at the site, not to mention the hundreds of waste shipments a year trundling along road or rail links to the repository. And how confident can anyone be that groundwater will not at some point invade the repository, corrode the waste packages, leach out the radionuclides, and transport them toward potable water supplies? How much is really known about groundwater behavior over the immensely long time periods at issue? Isn't it possible that climate changes over such periods will greatly increase the rate of water infiltration into the repository? How much is known about the corrosion and leaching behavior of the materials used to immobilize the radionuclides in the waste containers? What about the risk of seismic or volcanic activity over these long time periods? And even if the area surrounding the repository is sparsely populated today, how can we be sure that it will remain so in the future? Isn't there a danger of a future mineral prospector stumbling on the site? Even worse, might some future nuclear felon intentionally reenter the repository in an attempt to recover the large inventories of weapons-usable plutonium that will be buried there (if spent fuel is disposed of directly)?

In the face of such alarming scenarios, the inclination of the environmentalist is to oppose the siting and construction of a repository. The only question is how strongly. The environmental community is divided on this point. To some activists, the problems loom so large that until they are resolved the only responsible course of action is to shut down all operating nuclear reactors. Others, worried about the environmental consequences of increased reliance on fossil fuels, are pulled in opposing directions, lacking confidence in nuclear waste repository technology, but unwilling to advocate a complete shutdown of the nuclear power industry. They may be willing to support, or at least tolerate, the "least worst" temporary measure – such as an interim storage facility for spent fuel – while alternative approaches to disposal are explored. Still other environmentalists, adamantly opposed to nuclear energy in any form, see the nuclear waste problem as the Achilles heel of the whole industry. For them, opposition to a repository is an effective tool for forcing the abandonment of the nuclear option, since the accumulation of waste, with no prospect of being able to dispose of it, provides one of the most powerful rationales for shutting down all nuclear power plants.

From elected officials we can neither expect the system optimizing behavior of the engineers, nor, in general, the environmentalists' pursuit of a single overriding goal. For elected officials, the problem of nuclear waste management is defined first and foremost in terms of the interests of their constituents. These vary widely. Politicians from potential repository "host" regions are likely to oppose efforts to site a repository there, or at the very least to try to accumulate enough influence to be able to block such efforts in the future. In contrast, politicians in areas that are heavily dependent on nuclear power are more likely to push for a quick solution to the waste disposal problem. Nuclear waste is the kind of problem that most politicians dislike intensely because any action is almost guaranteed to make many people very unhappy.

THE U.S. HIGH-LEVEL WASTE PROGRAM

How has the nuclear waste problem been dealt with in practice? How have these different interests and approaches intersected, and with what results? We focus here mainly on the U.S. program, but later draw comparisons with other countries' experiences. We consider three different aspects of the program: The repository site selection process itself; the issue of whether to construct an interim storage facility for spent fuel; and the regulatory standards that are to be applied to the repository. In each area, progress has been extraordinarily slow, despite several decades of effort. Taken as a whole, it is a record that few would regard as successful. Why?

Repository Site Selection

For more than a decade the government's high-level waste repository site selection program has focused exclusively on Yucca Mountain, a desert ridge in southwestern Nevada about 100 mi northwest of Las Vegas (see Figure 8.4). Political officials and others in Nevada have been strongly opposed to the Yucca Mountain project, and have repeatedly tried to prevent the Department of Energy from proceeding with it. Despite this opposition, President George W. Bush formally recommended to Congress in 2002 that Yucca Mountain be developed as the site for the nation's first high-level waste repository. The President's recommendation followed a large-scale scientific and engineering investigation of Yucca Mountain, culminating in a determination by the Department of Energy that the site was technically suitable for development. The DOE will next submit an application for a construction license to the federal Nuclear Regulatory Commission. If the necessary permits are granted, the official schedule calls for the facility to be available to receive waste beginning in 2010. Opposition remains strong in Nevada, however. Whether the state can legally block the repository altogether is doubtful, but as a practical matter state officials and residents are capable of causing long delays in the schedule, and privately even the most optimistic DOE officials do not expect a repository at Yucca Mountain to begin operating much before the year 2020 – half a century after the federal government first began seriously to search for a repository site. In the meantime, nuclear power has gone from being the nation's most promising new energy source to an industry in gradual decline, and the lack of progress in siting a high-level waste repository is one of the most widely cited causes of this reversal. What went wrong?

One possibility is that the failure is technical in origin. Perhaps the basic technology of deep geological disposal itself is fundamentally flawed. The weight of evidence indicates otherwise, however. A succession of technical assessments, some conducted by government agencies and others by independent groups of leading scientists and engineers, have all concluded that the geologic disposal approach is in principle capable of safely isolating the waste from the biosphere for as long as it poses significant risks. These studies agree that there are extensive geohydrological regions of the continental United States with the characteristics necessary to assure long-term isolation. Studies conducted in other countries have reached similar conclusions.

Another hypothesis is that the U.S. high-level waste program has suffered from insufficient funding. But this, too, turns out not to be correct. The program has had adequate financial resources at least since the early 1980s, when Congress enacted legislation setting up a financing mechanism under which each nuclear utility is required to deposit into the interest-bearing federal Nuclear Waste Fund 0.1¢ for every kwhr of electricity that its nuclear power plants generate. This Fund pays for the Department of Energy's nuclear waste program activities. Between 1983 and 2000 the utilities paid $10.5 billion into the Fund, and even though the Department of Energy has spent $5.5 billion from the Fund – more than half of it on the Yucca Mountain project – the Fund balance in 2000 was still nearly $10.5 billion (a testament to its interest-earning power). The 0.1¢ fee is periodically evaluated to determine whether the Waste Fund will be large enough to pay for the entire lifecycle cost of the waste repository program, including site characterization, facility construction and operation, and the decommissioning and post-closure monitoring of the repository. On each occasion so far the fee has been found to be adequate.[1]

The real reasons for the delay in establishing a repository are neither strictly technical nor economic. One factor is the strong local opposition that the prospect of a repository has invariably aroused, along with the ample opportunities afforded by the nation's legal and political institutions for disaffected local citizens and their representatives to resist federal initiatives. Another key reason is that federal nuclear operations have been plagued by a long history of missteps and mismanagement. The full extent of these problems often did not come to light until much later, because of government-imposed secrecy restrictions, especially in the nuclear defense sector. (The leaks from the defense high-level waste tanks at Hanford and Savannah River are an example.) The cumulative effect has been to undermine the credibility of the Department of Energy in the eyes of many whose acquiescence or agreement the agency needs in order to proceed.

The circumstances under which the federal government can override state and local wishes in the larger national interest is a question that goes to the heart of the U.S. constitution, and the issue has loomed large in every congressional attempt to enact

---

[1] For the latest assessment, see http://www.rw.doe.gov/techrep/feead_toc/feeadr14.pdf.

nuclear waste legislation. Though the precise details have varied, over time Congress has moved to grant a potential host state the right to veto a federal decision to site a repository in its territory, while retaining for itself the authority to override such a veto by majority vote. In crafting this policy, legislators have sought a balance between what they see as the federal government's constitutional prerogative and the political reality that a state that is opposed to a repository will be able to delay it significantly, and perhaps even indefinitely. As a practical matter, the chances that a federal repository siting initiative will succeed in the absence of significant political support at the state and local level are not high.

The government has tried various approaches to securing local support, but none has been successful. An early strategy was to conduct a systematic, nationwide search for the "best" location. The 1982 Nuclear Waste Policy Act – the first comprehensive piece of federal legislation to address the high-level waste problem – called for an elaborate screening process covering the entire lower 48 states, which would eventually lead to the identification of two sites, one west of the Mississippi and the other, five years later, in the East. The systematic screening methodology was intended to produce a technically defensible result that would also meet the anticipated public demand for a fair process. The two-repository requirement addressed the fairness issue in a different way. On technical grounds there was no reason why a single repository could not store all of the nation's waste. But the most likely location for such a site would be the arid, sparsely populated West (which was where most of the geological investigations had focused previously). Yet most of the nation's nuclear power plants are in the East. The two-site strategy was the product of a political compromise. Without it the legislation would not have gained the support of western legislators, and without this support the legislation could not have passed.

The subsequent search for a site in the East turned out to be so controversial, however, that in 1985 the Reagan administration, faced with strong opposition from politically powerful eastern and midwestern senators and congressmen, abandoned it. The effect of this decision was to destabilize the fragile congressional coalition that had enabled the passage of the 1982 legislation, and before long the site selection process ground to a halt. In 1987 the Congress changed course. Abandoning any pretense of a "best-site" strategy, it directed the DOE to focus all of its resources on characterizing a single location – the Yucca Mountain site. Although Yucca Mountain had been one of the sites under investigation in the West, it had not emerged as the best of the candidates, and there is no doubt that politics played a large role in its selection. The small and politically weak Nevada congressional delegation, although opposed to the choice, was unable to muster the votes to prevent it. The DOE subsequently sought to gain local support for Yucca Mountain, but fifteen years later opposition remains strong. None of the leading elected officials in Nevada supports the project, and public opinion in the state is such that no candidate for public office can afford not to oppose it.

The DOE, years behind schedule despite its multibillion dollar outlay on studies of Yucca Mountain, is under heavy pressure from the nuclear utilities and others to move faster. But the Department is intensely aware of the importance of establishing and maintaining the scientific legitimacy of its work at the site. Especially given the management problems that have repeatedly surfaced elsewhere in the federal nuclear complex, DOE officials know that a perception of biased or shoddy technical decision-making could be fatal to the project. Many in Nevada believe, however, that the process has been unfair from the beginning. They point out that the federal government has no backup to Yucca Mountain and are convinced that an eventual decision by the Nuclear Regulatory Commission to issue a license to build a repository at the site is a foregone conclusion. Even the most rigorous technical justification of such a decision is unlikely to win them over.

The controversy over Yucca Mountain is thus all but certain to continue, and further delays seem inevitable unless the Federal government finds a way to generate significant local support for the project. With no such strategy in prospect, and with no backup plan either, the government's position is less than secure.

Interim Storage of Spent Fuel

With the opening of a repository still many years away, temporary storage capacity for spent fuel continues to grow scarcer. Many utilities have been able to expand their on-site storage capacity, but space constraints and regulatory obstacles have prevented others from doing so, raising the possibility that their reactors will be forced to shut down prematurely. Some reactors have already been closed for other reasons, but their owners cannot free up the sites for other uses without moving the spent fuel away.

From a systems engineering perspective, it may seem odd that this problem has not been solved. If policymakers had realized at the outset of the nation's nuclear power program that it was going to take several decades to locate and build a high-level waste repository, and that spent fuel reprocessing would be abandoned in the meantime, they would surely have concluded that interim storage arrangements for the spent fuel would be necessary. Of course, they, unlike us, did not have the benefit of hindsight. Still, as the delays and policy shifts took hold, the systems engineering case for such a scheme became more and more compelling. Why was nothing done?

There have in fact been several attempts to establish an interim spent fuel storage facility, all unsuccessful, and an account of why these failed reveals the limitations of a purely systems engineering perspective. Finding a site for a temporary storage facility has proved almost as difficult as locating a permanent repository. The task has been further complicated by disagreements over who is responsible and who should pay for it – the utilities or the federal government – as well as what the role of such a facility should be. Finally, the question of interim spent fuel storage has been caught up in the broader political dispute over the future of nuclear power. From a systems engineering perspective the two issues ought to have been separable, but in practice approaches to the spent fuel storage question have been strongly colored by positions on the larger issue of nuclear power's future.

The delay in establishing a permanent geologic repository, while increasing the pressure for an interim storage facility, has simultaneously complicated its prospects, by adding to fears that it will become a *de facto* permanent site. Some experts have in fact argued that an engineered surface or near-surface storage facility should be built as an alternative to geologic disposal for the next 100 years or so. They assert that much more will be known about geologic disposal by then, that the controversy over siting a repository will probably have faded, and that the spent fuel can be safely contained in an engineered structure in the meantime. For a brief period during the 1970s the U.S. Atomic Energy Commission and its short-lived successor, the Energy Research and Development Administration (soon to evolve into the Department of Energy) actually advocated this approach, but the policy was quickly abandoned in the face of pressure for a permanent solution.

The interim storage issue re-emerged in the late 1970s, however, following the Carter administration's decision to defer spent fuel reprocessing indefinitely. With no possibility of shipping their spent fuel elsewhere, the nuclear utilities urged the government to supplement their on-site storage pools with a centralized storage facility. But little progress was made, partly because the government took the position that moving forward with interim fuel storage would delay progress on permanent disposal.

As the repository site selection process faltered during the 1980s, the question of temporary fuel storage grew more urgent. Legislation enacted in 1987 authorized the government to build a monitored retrievable storage facility (MRS) for the fuel. But permission to proceed with construction was made contingent on the issuance of a construction license for the geologic repository, and the legislation also imposed a capacity limit on the MRS of 15,000 tons – much less than the total expected spent fuel inventory of more than 75,000 tons. Both provisions were designed to prevent the MRS from becoming a *de facto* alternative to a final repository. The legislation also explicitly prohibited the MRS from being located in Nevada – a propitiatory offering to the state's delegation, whose strong opposition to the Yucca Mountain project had been overridden in the same legislation. Later, however, as pressure for an MRS mounted, new legislation was introduced authorizing construction of a temporary storage facility at Nevada's National Nuclear Weapons Testing Station, not far from Yucca Mountain. Congress passed this bill in 2000, once again in the face of strong protests by Nevada, but President Clinton vetoed it on the grounds that an MRS in Nevada would undermine public confidence in the scientific legitimacy of the Yucca Mountain project. The President was also mindful that an MRS could hurt the Democratic candidate's prospects in a tight Nevada senatorial election race. As expected, the Congress did not override the President's veto, and the question of interim spent fuel storage remains unresolved.

The failure to resolve this issue has been a long-running source of tension between the nuclear industry and the federal government, and has exposed the government to large damage claims. The 1982 Nuclear Waste Policy Act had directed the DOE to sign contracts with every nuclear utility under which the Department was obligated to begin accepting spent fuel for final disposal in 1998 (by which time the Congress expected the first geologic repository to be available). When 1998 arrived with an operating repository nowhere in sight, the DOE sought to renege on this commitment, and a number of utilities sued the Department for breach of contract. The courts found DOE to be financially liable, and more than a dozen utilities are seeking monetary damages which may eventually total billions of dollars. The court's judgment was influenced by the fact that the utilities had already paid billions of dollars into the Nuclear Waste Fund, supposedly to enable DOE to take care of the problem. The DOE has since proposed that it should take ownership and management responsibility for spent fuel at reactor sites and pay for temporary storage if necessary – reversing the longstanding policy that this should be the utilities' responsibility.

Regulation of High-level Waste Repositories

The development of a regulatory framework is a third aspect of the U.S. high-level waste program where progress has been extraordinarily slow. The regulator of a high-level waste repository confronts two fundamental questions. First, what standard of safety performance must the repository achieve? In other words, how safe is safe enough? Second, how to determine whether the repository will be in compliance with this standard? Absolute certainty is obviously impossible over the very long time scales involved, but before granting permission to proceed the regulator must be reasonably confident that the safety standard will be met.

These questions present difficult problems for scientists and engineers, but they also raise issues extending far beyond the technological domain. The question of "How safe is safe enough?" is hard enough to deal with when it concerns today's society, but is harder still when the health and habits of distant generations are at issue. Similarly, the problem of verifying compliance raises the question of what constitutes an acceptable level of confidence. Questions of this type have been called "transscientific." They can be formulated in the language of science, but they cannot be answered by scientific methods alone.

These difficult issues have been further complicated by bureaucratic wrangling between different regulatory agencies with overlapping jurisdictions. Under U.S. law, regulatory authority over high-level waste disposal rests primarily with the federal government. (The federal government, through the Department of Energy, is also responsible for all aspects of building and operating high-level waste repositories.) Although several federal agencies claim some regulatory jurisdiction – for example, the Department of Transportation over the transport of radioactive materials – the two agencies with principal authority are the Environmental Protection Agency (EPA) and the Nuclear Regulatory Commission (NRC). The EPA is responsible for determining the general standards for protection of human health and the environment that the repository must meet, whereas the NRC is responsible for establishing specific technical criteria for the repository. The NRC also has licensing responsibility for the facility, and is required to issue construction and operating licenses to the DOE once it is satisfied that the facility will meet both its own and the EPA's performance criteria. The jurisdictional boundaries between the EPA and the NRC partially overlap, and the two agencies have frequently been in conflict. The Congress, frustrated by the slow progress and internecine warfare, has periodically attempted to insert itself directly into the technical standard-setting process, and has also involved the National Academy of Sciences in an effort to ensure that the outcome is scientifically sound.

How Safe is Safe Enough? One of the areas of dispute concerns the magnitude of the risk that the repository should pose to individuals living nearby (the question of "how safe is safe enough.") The EPA has stipulated that the dose to the maximally-exposed individual living close to the site should not exceed 15 millirems per year during the first 10,000 years after waste emplacement. This standard has been criticized as both too low and too high. (At one point the Nuclear Regulatory Commission proposed a 25 millirem per year standard, while legislation introduced into the House of Representatives called for a 100 millirem per year standard.) It is instructive to compare these dose limits with the average dose of about 300 millirems that individuals receive annually from natural background radiation, mainly from cosmic rays and terrestrial sources (including radioactive potassium in the thyroid). The maximum dose of 15 millirems per year proposed by the EPA translates into an annual risk of developing a fatal cancer of about 1 chance in 100,000. In setting this limit, the EPA also took note of the recommendation from the International Commission on Radiological Protection that the dose to members of the public from all sources of radiation other than background and medical applications should not exceed 100 millirems per year.

Any attempt to prescribe the performance of a waste repository must recognize the high likelihood that social behavior and technological capabilities will change radically over such long periods. Some critics have taken issue with the whole notion of setting quantitative limits because of this, but the EPA adopted a different approach:

> To avoid unsupportable speculation regarding human activities and conditions, we believe it is appropriate to assume that . . . parameters describing human activities and interactions with the repository (for example, the level of human knowledge and technical capability, human physiology and nutritional needs, general lifestyles and food consumption patterns of the population, and potential pathways through the biosphere leading to radiation exposure or humans) *will remain as they are today.*[2] (emphasis added)

Verifying Compliance. No less problematic is the issue of verifying repository compliance with the regulations. The physical size and intrinsic variability of the repository's geologic and hydrologic barriers mean that they cannot be characterized as fully or as precisely as manmade components. Moreover, neither they nor the engineered barriers can be tested over their projected lifetimes in the manner that is customary for the components of engineered systems such as aircraft or nuclear power plants. To a greater degree than for those systems, the determination of whether a repository can meet the safety standard will depend on expert technical judgments that will not be "provable" in a strict scientific sense. The credibility of those technical judgments will thus be a key

[2] See U.S. Environmental Protection Agency, "Public Health and Environmental Protection Radiation Standards for Yucca Mountain, Nevada: Final Rule," 40 CFR Part 197, issued June 5, 2001, http://www.epa.gov/radiation/yucca/docs/yucca_mtn_standards_060501.pdf.

issue. It is important to recognize that credibility will not be solely or even primarily determined by the reputations of the experts among their peers in the technical community. Regulation does not take place in a vacuum. The repository project, already the focus of intense public scrutiny, will attract still more as it proceeds, and in this environment, the credibility of regulatory judgments will be strongly affected by the level of public trust and confidence in the regulators, as well as in those they are regulating. A critical task for those responsible for the waste repository program, therefore, must be to try to create a climate in which the public trust in the technical community can flourish.

INTERNATIONAL COMPARISONS

As Table 8.1 showed, the long lead-time required to develop a high-level waste repository is not unique to the United States, and several other countries have also experienced strong local opposition to exploration activities. In some respects, however, U.S. policy diverges significantly from the approach taken elsewhere. For example:

- In some countries, such as Canada and Sweden, enough spent fuel storage capacity has been established to meet the needs of the nuclear power reactors for their full operating life. This has reduced the pressure to open a repository.
- In most countries, more of the operational responsibilities for spent fuel storage and disposal has been assigned to those responsible for actually generating the waste – that is, the nuclear utilities. In Sweden and Switzerland, utility-financed organizations will build and operate the waste repository.
- Most countries' repository programs are not as schedule-driven as the U.S. program. Some have chosen to defer the search for a repository for several decades.
- Other countries have taken a less prescriptive approach to waste disposal regulation, establishing broad safety goals for the repository but leaving more of the detailed design decisions to the waste management organization.
- As previously noted, some countries have placed more emphasis on the role of engineered barriers in the repository system. For example, the Swedish design calls for the waste to be contained in thick-walled copper and steel canisters, which are expected to remain intact for a million years. The Swedish engineers believe that the public will have greater confidence in a repository in which each barrier has been designed to maximize the probability of waste containment.

GENERAL LESSONS

The case of high-level waste management underscores several important lessons for technology practitioners. *First*, the job of the engineer does not end with the technical design of the system. Even though the concept of deep geologic disposal of high-level waste is judged by most knowledgeable scientists and engineers to be technologically sound, it has so far proved impossible to implement. Implementation is an integral part of technological practice in every field. Even the best technological solution creates no value to society until it is implemented. In this case, participating technologists must share in the responsibility for the failure to implement a waste repository system.

Second, technology practitioners cannot afford to externalize the social and political aspects of their work. For many years engineers and managers in the nuclear industry and in government downplayed the seriousness of the nuclear waste problem. They frequently declared that a technical solution was at hand, and that the obstacles were "only" political, implying that the latter were somehow unimportant. Later, as they came to gain a fuller appreciation of the scale and intensity of public opposition to repository initiatives, many concluded that the problem was mainly caused by a failure to communicate accurate technical information. Their proposed solution was to undertake public education and outreach activities – or, more precisely, to urge others to do so, because most did not see such work as an integral part of their own jobs. But public opposition did not only stem from a lack of accurate technical information. Deciding how to manage waste entails balancing different and sometimes conflicting values held by different groups and individuals in society. Moreover, much of the controversy has focused not on what should be done, but rather on who should decide – at root a political question. The response of many of the engineers and managers in the high-level waste program was to view these social and political issues as separate from their own work, something for others to deal with. But as we have seen, there is no possibility of implementing a technical solution unless these issues are considered. Instead of avoiding them, technical practitioners need to take account of these issues in their design work. They need to broaden their conception of the system they are working on.

Third, the engineers and managers who are responsible for the waste management program have learned that their credibility in the public domain is perhaps their most important asset that, once lost, is enormously difficult to rebuild. The cultivation and stewardship of the public trust and confidence is an integral part of sound technological practice. Once again, this is not something that can be left to others.

Much about the nuclear waste program is unique, of course, but the lessons from this case study have broader relevance. For technology practitioners in many other fields – genetically modified foods, biotechnology and health care, air transportation, and electronic commerce, to name only some of the most visible today – similar issues arise and similar lessons apply.

9

Nuclear Power and Weapons Proliferation

The discussion in the previous two chapters vividly illustrates the complexity of real world applications of nuclear reactor technology and its associated fuel cycle systems. We considered several issues that have brought this once promising technology to its knees: Economics, safety, and the environmental concerns surrounding waste management and disposal. In this chapter, we discuss the proliferation of nuclear weapons, another important problem related to the civilian use of nuclear power. Simply stated, commercial nuclear power carries with it a risk that technologies and materials from the nuclear fuel cycle will be misused for making nuclear bombs. There has been significant opposition to nuclear power on these grounds, and especially to nuclear exports to countries thought to be interested in acquiring nuclear weapons capability. In this chapter, we discuss the origin of these concerns, the history of how the proliferation issue arose, and the steps that have been taken by the international community to reduce the proliferation risks of nuclear power and, hence, to remove this obstacle to the peaceful application of nuclear technology.

Making Bombs

Nuclear fission weapons are made from either plutonium or highly enriched uranium (HEU).[1] When a relatively small quantity of either of these materials (on the order of 10 kg) is compressed by a modest amount of chemical explosives, an uncontrolled fission chain reaction can occur, releasing tremendous amounts of energy. A bomb weighing 100 kg can produce a nuclear explosion equivalent in power to the detonation of tens of thousands of tons of TNT (trinitrotoluene, a powerful conventional explosive). Moreover, a fission device can be used to compress light atoms such as tritium (a radioactive isotope of hydrogen containing two extra neutrons in its nucleus) to such high densities that they undergo nuclear fusion and release even greater amounts of energy. Thermonuclear fusion devices have the potential to produce explosions equivalent to tens of millions of tons of TNT.

There are three essential requirements for building a bomb: Possession of a sufficient quantity of plutonium or highly enriched uranium, knowledge of the principles of making a nuclear device, and considerable engineering effort to fabricate the device along with the fusing and delivery system necessary to make a credible weapon.

---

[1] HEU is enriched in excess of 80% ^{235}U. Recall that natural uranium contains about 0.7% of this isotope, and that commercial light water power reactors require low enriched uranium (LEU), containing 3% to 5% of ^{235}U. In principle it is possible to make nuclear fission devices with materials other than plutonium or uranium enriched with ^{235}U, but this is unusual.

Table 9.1. Routes to initial nuclear weapons

Plutonium route	HEU route
United States*	South Africa (aerodynamic enrichment)
Russia*	Pakistan (centrifuge enrichment)
China*	
United Kingdom*	
France*	
Israel	
India	
North Korea	

* Officially designated nuclear weapon states.

Since 1945, the knowledge of how to make a rudimentary fission nuclear device has spread quite widely. Many nations also today possess the engineering capability required to fabricate a credible weapon in a clandestine program. The main practical barrier is acquiring the plutonium or highly enriched uranium. Ten nations are known or believed to have acquired nuclear weapons capability. Eight of these nations made their initial weapons with plutonium, produced by irradiating uranium fuel in dedicated reactors. Pure plutonium metal was obtained by reprocessing the irradiated fuel assemblies containing the plutonium. Two nations acquired their initial capability with highly enriched uranium devices. The picture is summarized in Table 9.1.

Several other important points can also be made on the basis of previous experience:

1. Although, until now, no nation has employed plutonium derived from the commercial nuclear power industry for making weapons, the nuclear power fuel cycle is an obvious source. A single 1,000 MWe light water reactor produces about 250 kg of plutonium each year, which is enough for about 25 weapons. Of course, the plutonium is contained in the highly radioactive spent fuel and it becomes usable in a weapon only when it is separated by reprocessing the fuel. Thus, reprocessing technology and reprocessing plants are a key proliferation risk. More generally, ostensibly peaceful nuclear research and nuclear power programs can provide "cover" for purchases that are actually intended for weapons purposes, and for building relevant expertise and infrastructure.[2]

2. As of early 2002, there is no credible evidence that a nuclear weapon has been made by a sub-national group, although the possibility exists. For a sub-national group, the likely path to acquiring the necessary nuclear material is either by theft or purchase. This threat accounts for the tremendous importance of guarding all elements of the nuclear fuel cycle, especially reprocessing facilities.[3]

[2] The plutonium for India's first nuclear test in 1974, which was ostensibly a "peaceful nuclear explosion," was extracted from fuel from a research reactor that had been provided by Canada for civilian purposes.
[3] There are also concerns that sub-national groups may try to develop so-called "dirty bombs," in which radioactive materials are dispersed over wide areas using conventional explosives.

3. Testing is not required for a nation to have reasonable confidence that its first nuclear weapon will work. Several nations are understood to have acquired one or more nuclear weapons without testing, notably Israel, South Africa,[4] and, for many years, India and Pakistan.[5] Testing is certainly useful to "optimize" bomb designs, such as has been done by the five official nuclear weapon states, but a nuclear test is not necessary to have confidence that a crude bomb design will work, although perhaps with uncertain yield.

WHY DO NATIONS SEEK THE BOMB?

We know what motivated the five official nuclear weapons states to seek a nuclear capability. After the United States acquired the bomb to end World War II, the possession of nuclear weapons became the sine qua non of conducting the Cold War. The initial handful of nuclear weapons in the United States and the Soviet Union grew into the massive arsenals that each side insisted were needed for "deterrence." Each side sought to maintain a secure retaliatory capability, that is, a weapons arsenal considered sufficient to survive an unexpected first strike by the enemy and permit the state to retaliate in a manner that would inflict unacceptable damage on the attacking country. Deterrence was maintained by a balance of terror. Both the United States and the Soviet Union built and maintained tens of thousands of warheads. The nuclear weapons were not intended to be used, but rather were maintained in credible deployment to deter any possible attack. The United Kingdom, France, and China also sought deterrent capability, although on a much smaller scale. So much for the basic justification for nuclear weapons among the five official weapons states.

But why should smaller nations seek a nuclear capability? There is no doubt that some small countries have had this ambition. Today, North Korea, Iraq, and Iran are prominent examples. In the past, South Africa sought and successfully acquired a nuclear weapons capability, although it has since renounced the effort, and several nations – Taiwan, South Korea, Argentina, and Brazil – at various times have seemed to be interested in the possibility.

None of the newer members of the nuclear club – Israel, India, and Pakistan – have sufficient weapons to threaten seriously the major powers. These nations have acquired a nuclear capability because they believe possession of a bomb will help meet security concerns in their region, certainly in political terms and perhaps also in military terms. Israel has a (formally unacknowledged) nuclear capability because of the threat it perceives from its Arab neighbors. India and Pakistan have developed and deployed nuclear weapons as part of their ongoing conflict on the sub-continent. The possession of weapons by these countries has in turn influenced China's nuclear ambitions.

---

[4] South Africa abandoned its nuclear weapon program in 1991.
[5] India exploded its first device in 1974. In 1998, India conducted a series of tests that prompted Pakistan to reveal its capability with its own test series.

Diplomacy's Effect on Nuclear Power

Because of the terrible destructive power of any nuclear weapon, slowing the spread of nuclear weapons has been an important goal of international diplomacy ever since the atom bombs were dropped on Hiroshima and Nagasaki in 1945. This remains true today. Misuse of commercial nuclear power (for example by diverting separated plutonium from a commercial reprocessing plant, or by using a commercial enrichment facility to enrich low-enriched uranium to HEU) is one possible route to a bomb. Accordingly, the nonproliferation issue is yet another challenge that nuclear power must address.

The first attempts of U.S. policy to deal with the threat of nuclear proliferation date from the end of World War II. Already by then U.S. leaders clearly understood the danger of other nations acquiring nuclear weapons and the possibility that civilian nuclear activities could become a source of both weapons material and weapons-relevant technology. The United States sought a formula that would reduce both the incentives for and the ability of other countries to acquire nuclear weapons, while meeting the legitimate desire of many countries to participate in what was then considered to be an extraordinarily promising new energy technology. The solution found was the Eisenhower administration's "Atoms for Peace" policy, announced in 1953. This policy stipulated that, in exchange for nations agreeing not to pursue a bomb program, the United States would encourage the transfer of nuclear technology that would permit them to enjoy the fruits of peaceful uses of nuclear energy. Several countries, notably including India and other developing nations, saw this approach as discriminatory because it permitted great powers like the United States (and later other countries) to maintain nuclear weapons while forbidding other nations from acquiring them. These countries pressed the United States to agree to the unenforceable declaration that it would strive to achieve a world of total and complete disarmament of nuclear weapons.

For two decades "Atoms for Peace" was a central pillar of U.S. nuclear policy. Firms in the United States such as Westinghouse and General Electric worked to transfer nuclear technology and build nuclear power plants in Europe, Asia, and elsewhere. The U.S. government supported these power plant sales with long-term contracts to supply reactors with low-enriched uranium fuel. A prerequisite for U.S. nuclear exports was that there be in place a government-to-government agreement giving the United States the right to approve (or disapprove) subsequent re-transfers of nuclear materials and technology of U.S. origin to third countries. Thus, for example, a country wishing to transfer spent power reactor fuel that had been enriched in the United States to another country for reprocessing would have to gain the approval of the U.S. government first. Other rules governing nuclear transfers from the United States and other countries were designed and implemented by the Vienna-based International Atomic Energy Agency (IAEA), which had been established with strong United States support in 1957. These rules included the application of IAEA safeguards, including inspections of specified nuclear facilities, to verify that unauthorized diversion of nuclear materials or other inappropriate uses were not occurring. In hindsight, some of the rules seem decidedly odd. For example, only nuclear facilities that were actually declared by the host state could be inspected.[6] Also, a category of "peaceful nuclear explosives" (PNEs) was invented to distinguish military uses of nuclear explosives from possible civilian uses for, say, mining, although any PNE would, with a suitable delivery mechanism, be a bomb. Both the United States and Russia carried out so-called peaceful nuclear explosions in the 1960s and 1970s.

The most important nonproliferation policy accomplishment of the period was the entry into force in 1970 of the Treaty on the Nonproliferation of Nuclear Weapons (the NPT), following ratification by 43 countries. This treaty once again balanced the agreement of nonnuclear weapons states to renounce nuclear weapons with a commitment on the part of the five official nuclear weapons states to reduce and eventually eliminate their nuclear arsenals and to help other states realize the benefits of peaceful nuclear energy. The basic provisions of the NPT required the nonnuclear weapons state signatories to agree not to acquire nuclear weapons and to accept IAEA safeguards on all their civilian nuclear activities. For their part, the nuclear weapons states agreed not to provide nuclear weapons technology to nonnuclear weapons states, and to negotiate in good faith toward general and complete nuclear disarmament. Finally, all parties agreed to cooperate in peaceful uses of nuclear energy. The NPT, by obligating the nonnuclear weapons states to accept IAEA safeguards on all their facilities (instead of just certain of those that had been supplied in international commerce) greatly strengthened the IAEA safeguards regime. Even more important, the NPT helped to strengthen international norms against the spread of nuclear weapons. The NPT could not have been negotiated without the cooperation of the United States and the Soviet Union. Although the two Cold War adversaries agreed on little else, they shared the goal of preventing nuclear weapons spread, and the Soviet Union's support at key points was crucial to the treaty's eventual acceptance.

Everything changed in 1974 when India detonated a nuclear explosion in the Rajasthan desert. Until that time, optimism had prevailed about the possibility of preventing the spread of nuclear weapons. However, the Indian nuclear explosion caused an immediate re-examination of nonproliferation policies and what was happening in the world. There were two important observations. First, international commerce in nuclear power was increasing rapidly. The expansion went beyond power reactors to encompass the entire fuel cycle, including both uranium enrichment and the reprocessing of spent fuel.[7] At the time, Germany was planning to sell enrichment and reprocessing plants to Brazil, France was preparing to sell reprocessing plants to Pakistan and South Korea, and U.S. firms were planning to sell elements of the fuel cycle to South Korea, Japan, and Taiwan. Clearly, if these arrangements went ahead unchecked there would

---

[6] The most famous example of what could happen under such rules involves Iraq. Iraq permitted IAEA inspectors to inspect its OSIRIS research reactor. On their way to the declared reactor, the international inspectors would walk without question through an enormous facility that actually housed the extensive (but undeclared) Iraqi nuclear weapons program. The inspectors dutifully inspected OSIRIS without incident, concluding that Iraq was in compliance with its obligations.

[7] In that time of optimism about the future of nuclear energy, the separated plutonium was expected to be used first in mixed-oxide fuel in light water reactors and later in liquid metal fast breeder reactors.

be a major increase in the amount of sensitive technology and weapons usable material throughout the world.

The second observation was the increasing interest in nuclear weapons capability in many countries. Several countries were believed to have nuclear weapons programs underway or in the planning stages, including Taiwan, South Korea, Argentina, Brazil, and Pakistan. Most importantly, it was recognized that countries could move toward acquiring a bomb capability without mounting a covert or complete weapons program. Important steps would include developing a robust nuclear power industry that later might be a source of material and technology, as well as undertaking limited weapons design work.

The U.S. government began to oppose aspects of nuclear commerce. Beginning at the end of the Ford administration and continuing with increasing strength in the Carter administration, U.S. policy sought to discourage the reprocessing of commercial reactor fuel, oppose fast breeder reactor development (because these reactors use plutonium fuel), and restrict the development of new enrichment technologies such as laser isotope separation.

The nonproliferation concerns that arose in the United States in the aftermath of the Indian explosion were widely shared. Three groups were prominent: Defense specialists who saw the spread of nuclear weapons as a threat to United States and world security; environmentalists and other antinuclear groups who saw another reason to oppose what they already regarded as an excessively dangerous technology; and many academics who worried about the impact of the spread of nuclear weapons on international stability and arms control. It was an unusual and politically powerful coalition. Many individuals with these concerns entered the Carter administration and held high-level posts in the Departments of State, Defense, and Energy, as well as the National Security Council and the Council on Environmental Quality.

The shift in U.S. policy came as quite a shock to the U.S. nuclear and utility industries, which until then had enjoyed the support of the government in their pursuit of reprocessing, plutonium recycling, and fast breeder reactor development. The critics of the new policy in industry found powerful support in Congress, where some members were particularly concerned about the threat of cancellation of large nuclear projects that the government had previously supported. The impact of the policy shift on U.S. allies was even more pronounced. None of the important nuclear exporting countries, notably the United Kingdom, France, and Germany, had given much attention to proliferation issues. Their politicians and press were not prepared to deal with the technical complexity of the issue. As the United States invoked its right of approval of the re-transfer of nuclear material and technology of U.S. origin to third countries to slow the spread of sensitive technologies and plutonium use, political tempers flared. Policymakers in Europe and Japan did not understand the basis for the U.S. policy shift and suspected that the United States was trying to gain commercial advantage by the move. They contrasted their need for nuclear electricity with the ability of the United States, with its vast reserves of fossil fuel, to get by without nuclear power. They also argued that the new U.S. policy was inconsistent with its obligations under the NPT, according to which the advanced nuclear powers were expected to help other states develop nuclear energy, and that the United States was unilaterally seeking to proscribe certain civilian nuclear fuel cycle activities that were expressly permitted by the treaty. Matters were not helped by the inexperience of the new administration's technocrats, whose initial actions were viewed abroad as high-handed and poorly explained. The proliferation issue became the most contentious security issue for the Carter administration in its first years in office.

HOW INTERNATIONAL CONSENSUS WAS REACHED

Remarkably, however, within two or three years a working consensus was reached about nonproliferation. Progress came about through a long process of explaining the complex connection between nuclear technology and security. The explanation emerged from the public analysis of the consequences of unfettered nuclear commerce, along with discussion of alternatives that employed different fuel cycles and procedures that would reduce the risk of nuclear weapons spread.

On the domestic front, members of the administration with strong technical and security credentials carried the argument, so that their credibility was high with both Congress and industry. The prospects for success were enhanced when it became apparent that the main dangers in nuclear commerce – reprocessing, fast breeder reactors, and new enrichment technologies – were not likely to become economically competitive with the once-though fuel cycle for light water reactors for many decades. Many studies were done to establish this point, including studies of the availability of uranium ore, and of the cost of reprocessing and of fabricating mixed-oxide fuel. Environmental groups insisted that such studies be a required part of the licensing process of the Nuclear Regulatory Commission.[8]

Internationally, the administration led an initiative to establish the two-year International Nuclear Fuel Cycle Evaluation (INFCE) study that brought together experts from a wide range of countries to develop a common understanding of the risks, benefits, and costs of alternative nuclear fuel cycles. At the end of the INFCE process there was surprising agreement among the nuclear exporting countries about the dangers of unconstrained nuclear technology exports and the measures required for more prudent use of the technology.

Several other actions were also taken beginning in the late seventies:

1. Exports of technology considered to be sensitive – especially reprocessing plants and enrichment facilities – were cancelled. The leading nuclear supplier countries formed an informal Nuclear Suppliers Group to monitor proposed international transactions in the nuclear field.

---

[8] An excellent example was the so-called GESMO (for Generic Environmental Statement on Mixed Oxide Fuel) proceeding of the NRC – a thorough review of the environmental impact of reprocessing and mixed oxide fuel relative to alternatives.

2. The United States took the lead in efforts to modify the large number of operating research reactors around the world (many of them previously supplied by U.S. industry) so they could operate on low-enriched rather than highly enriched fuel.
3. Diplomatic efforts were undertaken to convince several nations to abandon their attempts to acquire nuclear weapons technology. These efforts were effective in reversing programs in Argentina, Brazil, South Korea, and Taiwan.

The Carter administration also took the initiative to gain additional international support for the Nuclear Nonproliferation Treaty (NPT), which has continued to attract new members and today has nearly 190 parties (more than have signed the U.N. Charter.)

All of these efforts helped to strengthen the worldwide consensus on the control of commercial nuclear technology. But the results, of course, have not been universally successful. The IAEA safeguards system, even with the strengthening that occurred after the revelation of Iraq's nuclear capability, is not foolproof. Detection of a surreptitious bomb program can be very difficult.[9] The NPT did not prevent India and Pakistan (neither of them parties to the Treaty) from detonating nuclear weapons in a series of tests in the summer of 1998. On the other hand, the fact that Soviet nuclear weapons did not remain in Ukraine and the other states of the former Soviet Union but were all returned to Russia was a particular success of nonproliferation efforts in the early nineties.

PRESENT PROLIFERATION CONCERNS

Geopolitical circumstances have changed considerably in the last ten years. Most important, the Cold War is over and with it the nuclear balance of terror between East and West. The Soviet Union has collapsed, and thousands of nuclear weapons are no longer aimed at each other. The second change is that, as time has passed, technology and knowledge have inevitably spread. The result is growing concern over the spread of nuclear weapons and other weapons of mass destruction, not only to states such as Libya and Iran but also to terrorist groups.[10] The terrorist attacks on the United States on September 11, 2001 have intensified this concern. Accordingly, efforts to control technology transfer, especially for technologies that are "dual use" in nature (i.e., with possible military as well as commercial application), have intensified. Finally, the collapse of the Russian economy, beginning in the early 1990s, led to the weakening of administrative control over nuclear weapons throughout that country. Many facilities where nuclear weapons are stored and developed, or where nuclear materials are produced (through reprocessing or enrichment) and stored, no longer have the funds to provide for security or to pay guards and technical people. The quantities of weapons-usable

[9] In the late 1980s, North Korea managed to reprocess fuel and recover plutonium at its Yong-bon reprocessing plant before being detected. And in 2002, when confronted by the United States, North Korea admitted to having conducted a clandestine uranium enrichment program for several years. The North Koreans subsequently announced their withdrawal from the NPT, and forced the IAEA inspectors to leave the country.

[10] The other weapons of mass destruction are weaponized chemical and biological agents.

materials in Russian stockpiles are not known with certainty, but are thought to include 160 tons of plutonium and more than 1,000 tons of highly enriched uranium, as well as about 20,000 warheads.

For those seeking access to nuclear weapons or weapons-usable materials, the most obvious source is to divert them from the vast, undersupervised Russian stockpile. For a Russian nuclear laboratory that has no funds to pay its workers or to support the community in which the laboratory is embedded, the temptation to sell nuclear material or technology may be very great indeed.

Accordingly, for the past decade the United States has had a major initiative in place to control the "loose nukes" in Russia. First, the United States through the "cooperative threat reduction program," provides assistance to Russia to improve its stockpile stewardship.[11] Efforts have been made to install modern security and accounting systems at Russian facilities, to provide employment for Russian weapons scientists, and to convert or shut down plutonium production reactors. Second, the United States and its allies have introduced fissile materials programs to reduce the stockpiles of surplus highly enriched uranium and plutonium. The United States has agreed to buy 500 MTHM of Russian HEU that the Russians will blend down to LEU for use in commercial reactors. For plutonium, the United States and its allies have agreed to support a program that will convert about 35 MTHM of Russian plutonium into mixed oxide fuel for burning in Russian or European reactors. However, the cost of producing mixed oxide fuel is so large compared with the cost of LEU fuel, even if the military surplus plutonium is assumed to be available at zero cost, that the alternative of long-term plutonium storage under international auspices is also being explored. (Environmentalists who are opposed to any use of plutonium for energy generation have been strong advocates of the latter option.) Despite the considerable efforts that have been made in Russia, and the high level of cooperation between Russian and American experts, much more remains to be done.

In sum, over the past quarter century the connection between nuclear power and national security has been identified and confronted. The strategies for managing this link have taken four forms: (1) maintaining secrecy and restrictions on access to sensitive technologies and materials; (2) building international norms against the spread and use of nuclear weapons; (3) reducing national incentives to acquire nuclear weapons; and (4) promoting international cooperation in the peaceful uses of nuclear technology. Over the years the balance between these strategies has changed. None can be effective on its own. In recent years, the focus of nonproliferation efforts has shifted from misuse of the commercial nuclear fuel cycle – a possibility that was increased by the early "Atoms for Peace" policy – to containing "loose nukes" in Russia. Ignorance about nuclear technology in diplomatic circles has been replaced by a more sophisticated awareness of the technology and of what policy measures will work to reduce the risk of the spread of nuclear weapons. International cooperation among supplier nations is clearly crucial

---

[11] The program was established by legislation sponsored by Senators Nunn, Lugar, and Domenici, who have given bipartisan support to the effort.

to the success of this effort. For the nuclear industry, shock at being linked to nuclear weapons has been replaced by acceptance that it must take into account legitimate security concerns.

The underlying driving forces for the acquisition of nuclear weapons are unrelated to civilian nuclear energy – security concerns, the desire for international prestige, domestic political and bureaucratic factors, or, in the case of terrorist groups, a drive to inflict damage on societies that are perceived to be thwarting their religious or political aims. But civilian nuclear power programs can in some circumstances contribute to nuclear proliferation, and these risks must be minimized.

WHAT CAN WE LEARN FROM THE HISTORY OF NONPROLIFERATION POLICY?

As we have seen, the introduction of any technology has the potential to affect society in ways that were unintended by its developers. We have stressed the economic, environmental, and social impacts of new technology in this book, but in the case of nuclear energy technology an additional impact emerges – national security. While the link between nuclear energy and nuclear weapons is a particularly dramatic case in point, many other technology applications also have implications for national security, including the export of supercomputer, satellite, and encryption technologies, and the undertaking of major projects with potential adversaries, such as the construction of a pipeline from the Caspian Sea through Iran to the Persian Gulf.

Those who are engaged in introducing new technology in order to benefit society need, therefore, to consider the broadest possible range of impacts that the technology may have. Most importantly, if an unanticipated impact arises, proponents must be quick to respond to the new concerns. In the end, the nuclear industry responded relatively rapidly and constructively to proliferation concerns, but perhaps less so to the concerns about nuclear safety and radioactive waste that were discussed in the previous chapters.

10

Natural Gas

Natural gas will be the world's most important source of new primary energy supplies during the next two decades. Consumption of the fuel worldwide is projected to double by the year 2020. In the United States, natural gas is currently the second most important fuel after coal, accounting for almost a quarter of all energy consumed, and the U.S. Energy Information Administration projects that domestic consumption in 2020 will be 50% higher than today. Much of this growth will occur in the electricity sector. More than 90% of additions to U.S. electricity generating capacity over the next decade will be gas-fired. Natural gas is also an attractive residential and commercial heating source (most new single-family homes use gas heating) as well as an important industrial fuel.

The emergence of natural gas as the fuel of choice for electricity generation in the United States stands in sharp contrast to the situation less than two decades ago. After vigorous growth earlier in the twentieth century, U.S. gas consumption peaked in the early 1970s and continued to decline until the middle part of the 1980s (see Figure 10.1). During part of this period, the use of natural gas by electric utilities was actually prohibited by the federal government on the grounds that it should not be "wasted" by combustion in utility boilers but rather conserved for more important uses such as heating homes and as feedstock for chemical plants. In this chapter, we will examine the reasons for this striking reversal. Why has natural gas now become such an important part of the energy picture? Are today's optimistic projections realistic?

The Advantages of Natural Gas

The increasing emphasis on gas is attributable mainly to four developments:

- First, natural gas, which consists primarily of methane (CH_4), is a clean burning fuel, at least by comparison with other fossil fuels. Air pollutant emissions per unit of energy consumed are compared for natural gas, oil, and coal in Table 10.1. Natural gas emits far smaller quantities of nitrogen oxides, sulfur dioxide and particulates than does either coal or oil, and its CO_2 emissions per BTU are only 56% of those for coal and 70% of those for oil. The reason for the lower carbon emissions is that natural gas has the highest hydrogen/carbon ratio of all hydrocarbon fuels. This attribute may turn out to be particularly significant as part of an overall strategy for managing the greenhouse gas problem.

164

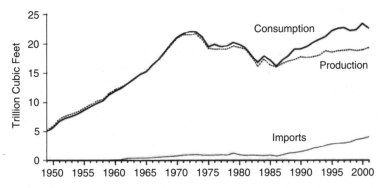

Source: Reprinted from Energy Information Administration, *Annual Energy Review – 1998*, www.eia.doe.gov/aer.

Figure 10.1. Trends in U.S. natural gas consumption and production.

- Second, natural gas is convenient to use in many applications, including electric power generation. It is easier to site natural gas-fired power plants than either coal or nuclear plants. And because natural gas is easier to handle and cleaner-burning than coal, requiring fewer costly pollution control systems, the capital and operating costs of gas-fired power plants are lower than those for coal. Construction lead times are also considerably shorter for gas-fired plants; a large gas-fired power plant can be designed and built in three or four years. In the uncertain environment of economic regulation now confronting many power generators, the lower capital cost and shorter construction lead-times of gas-fired power plants are especially attractive.
- Third, structural changes in the gas supply industry, stimulated by a long-term process of economic deregulation which began in the late 1970s, introduced more competition into the industry, lowered prices, and helped pave the way for an expansion of the network of pipelines used for long-distance transmission and local distribution of gas. The pipeline network had long been considered a natural monopoly, and the prices levied by pipeline operators were closely controlled by government regulators. Interstate gas pipelines were regulated by the federal government, while local distribution was regulated by state commissions. In 1973, when domestic gas prices increased sharply in response to the oil price increases caused by

Table 10.1. Pounds of air pollutants emitted per billion BTU of energy consumed (% in parentheses)

	Natural gas	Oil	Coal
Carbon Dioxide	117,000 (56)	164,000 (79)	208,000 (100)
Carbon Monoxide	40 (19)	33 (16)	208 (100)
Nitrogen Oxides	92 (20)	448 (98)	457 (100)
Sulfur Dioxide	0.6 (.02)	1,122 (43)	2,591 (100)
Particulates	7.0 (.3)	84 (3)	2,744 (100)

Source: Energy Information Administration, *Natural Gas 1998: Issues and Trends*, p. 53.

the OPEC oil embargo, regulatory controls were extended upstream to the price of gas that producers could charge at the wellhead. The goal was to protect consumers from sharply higher prices and to prevent the "windfall profits" that domestic gas producers would otherwise reap. But as the system of regulation grew more complex, it became increasingly unwieldy and difficult to implement efficiently. By the late 1970s, it was widely viewed as unworkable. Prices were in many cases held to unrealistically low levels, which encouraged high levels of consumption while discouraging exploratory drilling and new discoveries. Faced with an apparent shortage (despite the existence of large amounts of gas in the ground), the government curtailed gas use by some customers, while prices shot up in markets that were not price-controlled. Such gyrations contributed to a widespread perception of gas as a costly, scarce, and unreliable fuel. The result was that gas commanded a lower price at the "burner tip" than No. 2 fuel oil (diesel) – the competing fuel.

Starting with the Natural Gas Policy Act of 1978 and extending over more than a decade, the gas industry was gradually opened up to increased competition. The process began with the phased decontrol of natural gas prices at the wellhead. Also, large industrial users, who had previously been required to buy gas from the interstate pipeline companies, were permitted to contract directly with gas producers. The pipelines, which had previously held a monopoly over gas purchase and resale, were now required to provide transportation services for these large contracts. The pipeline companies were themselves allowed to compete with each other to supply the local distribution companies and, later, their end-use customers – a development that became possible because several gas pipelines now served the same area, so that more than one source of gas supply was available. In many states today, industrial and commercial customers as well as large residential customers can choose their own gas suppliers.

This competitive restructuring has been slow to develop and has often been controversial, but the net effect has been to attract more investment into the industry and to help create more confidence on the part of consumers that their demands for gas will be satisfied at reasonable prices.

Fourth, perceptions of the availability of natural gas in the U.S. have shifted from concerns over depletion and scarcity to a widespread belief that natural gas resources are relatively abundant. This may well be the biggest of the recent changes affecting the status of natural gas, and it raises two important sets of questions. First, how confident can we be in the resource estimates? How much gas is really available? Second, what should the nation's attitude be towards the use of this nonrenewable resource? How rapidly *should* it be consumed? These are fundamental questions not only for natural gas but for all depletable resources, whether non-renewable mineral resources or "semi-renewable" resources like forests and fish stocks. Conservationists worry that the world is consuming these resources too rapidly, and that today's generation is selfishly depriving future generations of a resource endowment that is rightfully theirs. Our discussion of natural gas in this chapter serves as a case study of how to think about this broader and very important topic.

Natural Gas Resources

As previously noted, natural gas consists primarily of methane (CH_4), but it may also contain lesser amounts of other low molecular weight hydrocarbons such as ethane (C_2H_6) and propane (C_3H_8). Pipeline quality gas typically contains 90% or more of methane. The gas occurs either in association with crude oil deposits or in free-standing ("nonassociated") gas reservoirs.

Natural gas resources are classified as either "conventional" or "unconventional." The conventional category includes gas, either unassociated or associated with oil, that occurs onshore or offshore up to depths of about 20,000 ft. In the United States, the majority of these conventional gas resources are located in the Gulf Coast region and in Texas.

Unconventional gas occurs in very deep deposits – there is no sharp dividing line separating deep gas from conventional gas deposits – as well as in "tight" sandstones, coal beds, and shale beds.

There are, in addition, vast deposits of natural gas hydrates – crystalline, ice-like substances that consist of a host lattice of water molecules containing molecules of methane and often small amounts of other gases. Natural gas hydrates are found mainly in permafrost regions like Siberia and Alaska and in marine sediments at water depths greater than 450 meters, where the pressure exceeds 45 atmospheres and the temperature is typically in the range of $4°C–6°C$. (At lower pressures and higher temperature natural gas hydrates are unstable.) The existence of these natural gas hydrate deposits has only been known since the mid-1960s, but huge volumes are believed to exist, with an estimated carbon content that dwarfs that of all other fossil hydrocarbons combined. In 1995, the U.S. Geological Survey estimated that U.S. deposits of natural gas hydrates contain about 320,000 trillion cubic feet (TCF) of methane. By comparison, "proved" natural gas reserves in the United States currently stand at less than 170 TCF. (We discuss the definition of these terms below.) None of this huge resource is economically viable at present, and no methods for extracting methane from natural gas hydrate deposits have yet been developed. But if even a tiny fraction of this resource turned out to be economically recoverable, the impact on natural gas supplies would be tremendous.

Geologists and resource economists use many different terms to describe and quantify the natural gas resource base. In addition to the "conventional" versus "unconventional" distinction mentioned earlier, other common descriptors include: "reserves" versus "resources;" "recoverable" versus "potential" resources; "technically recoverable" versus "economically recoverable" resources; "proved," "probable," "possible," and "inferred" reserves, and so on. Unfortunately there is no universally accepted definition for many of these categories, and in unpracticed hands their use can lead to inaccuracies in reporting and misunderstandings on the part of policymakers and the general public.

The diagram in Figure 10.2, the so-called McKelvey Box, helps to clarify matters. The McKelvey Box (named after a former director of the U.S. Geological Survey) displays the total resource base of any given mineral – that is, the entire amount of the resource that exists within the Earth – on a two-dimensional plot. Where any particular portion

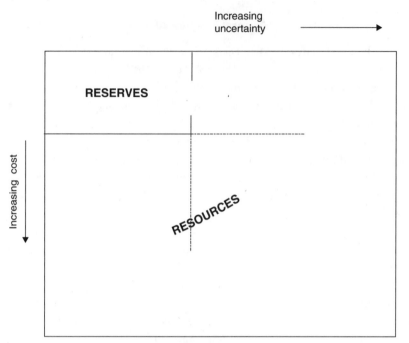

Figure 10.2. The McKelvey Box.

of this resource lies on the chart depends on two factors: The expected cost of extracting it, and how uncertain geologists are about whether it actually exists.

The upper left-hand corner of the diagram is occupied by deposits whose physical extent and quality (or grade) are well understood as a result of prior exploration and development activities, and whose cost of recovery makes them economically attractive under current market conditions. Moving out from this corner to the right along the uncertainty axis we encounter deposits that are less well-defined, and, still further to the right, other deposits that are only surmised to exist – for example in geologic structures which haven't been mapped but whose features are known to be similar to those of well-explored formations that are known to contain the resource. Similarly, as we move down the cost axis there are deposits that are uneconomic to recover at present prices, either because they are too deep, or not sufficiently high grade, or too difficult to process.

The lines on the McKelvey diagram dividing economic from uneconomic resources and known from unknown deposits are not sharp, but an important distinction is between *reserves*, which are well-defined and economic to recover and occupy the top-left region, and *resources*, which occupy the rest of the chart. The Energy Information Administration (EIA) defines reserves as "those volumes that are believed to be recoverable in the future from known deposits through the eventual application of present or anticipated technology." "Proved reserves," a still more restrictive category, are defined by the EIA as "those volumes that geological and engineering data demonstrate with reasonable certainty to be recoverable in future years from known reservoirs *under existing*

Table 10.2. U.S. natural gas reserves and resources

	Trillion cubic feet (TCF)
Proved Reserves	167.2
Discovered, Technically Recoverable Resources	
Reserves Growth[1] (Conventional; Onshore)	322.0
Reserves Growth[1] (Conventional; Federal Offshore)	32.7
Unproved Reserves (Federal Offshore)	5.1
Undiscovered, Technical Recoverable Resources	
Conventional (Onshore)	258.7
Conventional (Federal Offshore)	268.0
Continuous-type (in sandstone, shales, and chalks; onshore)	308.0
Continuous-type (in coalbeds; onshore)	49.9
Natural Gas Hydrates	322,222.0

[1] "Reserves Growth" is the volume by which the estimate of total recovery from a known oil or gas reservoir or aggregation of such reservoirs is expected to increase during the time between discovery and permanent abandonment.

Source: Energy Information Administration, *Annual Energy Review – 2000*, www.eia.doe.gov/aer (visited 1/17/02).

economic and operating conditions" (emphasis added). These are general definitions, applying to gas and oil resources and to other minerals too.

In the United States, the U.S. Geological Survey (USGS), an agency of the Interior Department, is the main source of official estimates of gas reserves and resources. (The Minerals Management Service, another Interior Department agency, is responsible for developing estimates of offshore gas resources in deeper waters under federal jurisdiction; offshore gas in near-shore and shallow water areas under state jurisdiction is estimated by USGS.) Independently, the Potential Gas Committee, an industry-sponsored group, also periodically produces estimates. The current resource picture is shown in Table 10.2. As is the case with almost all minerals, the estimate of resources is far larger than the estimate of reserves.

The McKelvey diagram in Figure 10.2 is superimposed on a constantly changing situation on (or in) the ground. Proven reserves are continually depleted over time by extraction and use, but this is offset, at least in part, by the creation of new reserves through ongoing exploration and development activities. Other deposits may move across the dividing line from resources to reserves as a result of changes in technology or in markets. If the price of gas rises, companies are willing to spend more on exploration and development, and at such times proven reserves tend to increase. Additions to reserves vary considerably from year to year due to expectations about future gas prices.

Technological innovation also has an important bearing on reserve estimates. Advances in exploration technology have made the search for new deposits more efficient, lowering the cost of adding to reserves. Figure 10.3 shows how the "finding costs" of U.S. onshore and offshore oil and gas reserves have declined since the early 1980s. Similarly, innovations in extraction technology have brought down the cost of production for many minerals, at a rate that has often outpaced the upward cost trend caused by the

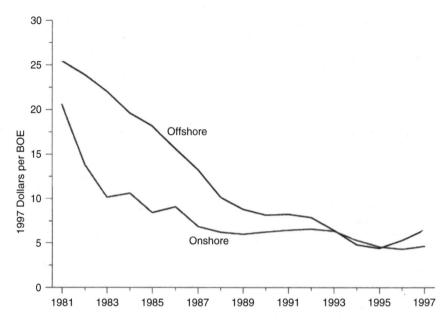

Note: Natural gas is converted to its oil equivalent using the conversion factor: 1 thousand cubic feet = 0.178 barrels of oil equivalent (BOE).

Source: Reprinted from Energy Information Administration, *Natural Gas 1998: Issues and Trends*, 101.

Figure 10.3. U.S. onshore and offshore finding costs for major energy companies, 1981–97.

decline in average deposit grade. For example, it costs less to mine a pound of copper today than it did fifty years ago, even though the average grade of copper ore mined today is much lower. Innovations in production technology have the effect of transforming deposits that were previously marginal or unattractive into economically viable reserves. In the gas industry, advances in offshore drilling technology have improved the economics of deepwater production and significant new reserves have been added in the deeper waters of the Gulf of Mexico as a result. Technological advances have also resulted in increases in the yield of gas from existing wells both on and offshore.

Recent trends in production rates, wellhead prices, and proven reserves are summarized in Table 10.3. A commonly cited indicator of the adequacy of reserves is the time it would take to deplete them at the current rate of production. As Table 10.3 shows, this depletion time has recently fallen to about eight and a half years, down from about twelve years in the mid-1980s. Yet, during this same period, as discussed earlier, expectations for the future role of gas have become more rather than less optimistic. What accounts for this seemingly contradictory situation? A large part of the explanation has to do with the improvements in exploration and production technology that occurred during this period, which had the effect of adding significantly to the estimates of resources that will eventually be recoverable economically. Some of these advances have made it possible to produce economically a larger fraction of the oil or gas in already discovered and partially depleted reservoirs.

Table 10.3. Production, price and reserves trends in the U.S.
natural gas industry

	U.S. natural gas production (trillion cubic feet per year)	Average well head price ($ per thousand cu. ft.)	U.S. proved natural gas reserves (billion cubic feet)
1977	19.2	.79	207,413
1978	19.1	.91	208,033
1979	19.7	1.18	200,997
1980	19.4	1.59	199,021
1981	19.2	1.98	201,730
1982	17.8	2.46	201,512
1983	16.1	2.59	200,247
1984	17.5	2.66	197,463
1985	16.5	2.51	193,369
1986	16.1	1.94	191,586
1987	16.6	1.67	187,211
1988	17.1	1.69	168,024
1989	17.3	1.69	167,116
1990	17.8	1.71	169,346
1991	17.7	1.64	167,062
1992	17.8	1.74	165,015
1993	18.1	2.04	162,415
1994	18.8	1.85	163,837
1995	18.6	1.55	165,146
1996	18.8	2.17	166,474
1997	18.9	2.32	167,223
1998	18.9	1.94	164,000
1999	18.6	2.17	167,400
2000	19.2	3.60	—

Source: U.S. Energy Information Administration, 2002.

Natural Gas Production

Most people would agree that it would be irresponsible for modern society to adopt a pattern of use of the earth's nonrenewable resources that would lead to their complete exhaustion. But behind this broad consensus there is a wide range of views concerning how much of these resources ought to be conserved for future generations, and how rapidly they should be exploited. An extreme minority would advocate prohibiting any use of nonrenewable resources whatsoever. At the other end of the spectrum are those who would dismiss all worries about resource exhaustion, on the grounds that new technology can always be relied upon to create alternatives (or make it possible to extract lower grade resources economically). In this view, apparently finite resources are in effect infinitely expandable, because human ingenuity will always find a way to trump natural constraints.

The prohibitionist view cannot be understood in economic terms. It is a philosophical or ideological position. The opposing view, in contrast, is rooted in the workings of the marketplace. It assumes that the scarcity value of the resource will be reflected in rising prices, which in turn will create powerful incentives for the development of substitutes for the depleting resource or new processes that will permit its extraction at lower cost.

The majority opinion lies somewhere between these two positions. Most people acknowledge the importance of market mechanisms for the efficient allocation of resources. At the same time, most people also recognize that ethical considerations ought not to be excluded from decisions concerning who should benefit from a fixed resource endowment, and in particular, what the intergenerational distribution of benefits should be.

It turns out that economic theory has something useful to say about both of these perspectives. The economic theory of exhaustible resources, first worked out by the economist Harold Hotelling nearly seventy years ago, predicts what will happen to prices and production over time in a competitive market for a finite resource.[1] As we shall see, this theoretical prediction isn't necessarily the socially optimal result, but it does provide a useful reference point against which to compare other scenarios.

The theory assumes that the owner of a gas reservoir (or copper mine, or any exhaustible resource – the theory is quite general) will seek to maximize the present value of his future profits. What pattern of production will achieve this?

Let us suppose that the profit the owner can realize on each unit of gas today, given the current market price and his cost of production, is π. (The profit is simply the market price less the unit cost of production, and is also referred to as the "scarcity rent" or the "net price.") The owner will produce gas today if he believes that he will earn more by selling the gas and investing the profits than by holding it in the ground and selling it next year. In other words, if the owner expects the market value of his gas to increase over time at a slower rate than the rate of interest, he will produce now. If, on the other hand, he believes that the value of his gas will rise faster than the rate of interest, he will hold production back. It follows that in a competitive market, the net price or scarcity rent, π, will increase exponentially at exactly the market rate of interest, r:

$$\pi(t) = \pi(0)e^{rt}.$$

(A formal derivation of this classic result is presented in the Appendix to this chapter.) If the price follows this trajectory, it will be a matter of indifference to the owner whether he produces gas today at price $\pi(0)$ or at time t at price $\pi(t)$.

Note that the market price seen by consumers will not necessarily exhibit the same exponential time trend, since the market price is equal to the sum of the scarcity rent and the production cost. If the production cost is declining over time the market price

---

[1] Harold Hotelling, "The Economics of Exhaustible Resources," *Journal of Political Economy*, 39(2), 137–175 (1931). For a valuable review of the theory, on which this account draws extensively, see Robert M. Solow, "The Economics of Resources or the Resources of Economics," *American Economic Review*, 64(2), 1–14 (1974).

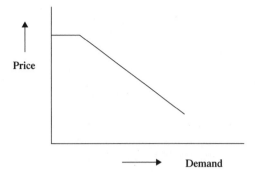

Figure 10.4. A demand curve for a nonrenewable resource.

might also decline, even though the scarcity rent is rising exponentially. (Whether this actually occurs or not depends on the relative magnitude of the rent and the production cost.) As pointed out previously, the market price of many nonrenewable resources has indeed continued to decline over the years, a fact often cited by those who argue, in opposition to the claims of many conservationists, that the world is not "running out" of resources in any meaningful sense.

Eventually, if the production cost continues to fall, the scarcity rent will come to dominate the movement of market price. Alternatively, the production cost may begin to rise as extraction becomes more difficult. In either case, the market price will eventually start to rise too, although this may not happen for a long time.

What does the theory have to say about the rate of production? Consider the demand curve for the resource shown in Figure 10.4. The demand curve describes the amount of resource that is consumed as a function of market price. In a competitive market the demand curve also determines the relationship between market price and production rate. At the current market price, production is just equal to demand and the market clears. As the market price increases, the demand (and therefore also the rate of production) will fall. Eventually there may come a point at which the price is so high that demand is choked off completely, and at that point production too will fall to zero. Assuming the market works efficiently, this will also be the point at which none of the resource is left in the ground.

The actual pattern of production over time will depend on the shape of the demand curve and the total supply of the resource. For the case of the simple linear demand curve shown in Figure 10.4 and a fixed reserve base, the theory predicts that the rate of production will decline steadily over time.[2]

In sum, the theory (which has been greatly simplified in this brief discussion) describes a dynamic competitive equilibrium: at each point in time the overall production

---

[2] In more complex – and more realistic – situations, producers aren't exploiting a fixed reserve base. Rather, they develop reserves through a (costly) process of exploration. The optimization problem for producers in this case involves deciding the optimal level of both production and exploration over time, and hence the optimal level of reserves as a function of time. In the early phases of a resource's life, producers typically are heavily involved in the discovery and development of reserves. Production is initially relatively low, but builds gradually over time. (See Robert S. Pindyck, *J. of Political Economy*, 86, 840 (1978).)

rate is the aggregation of the production decisions of individual producers, each of whom is producing at just the rate at which he is indifferent at the margin between producing and holding, with the current price equal to the market clearing price, and the scarcity rent or net price expected to increase over time at a rate equal to the rate of interest.

In a situation in which there are multiple resource deposits, each with a different cost of production, the theory predicts that the lowest-cost source will be used first. The market price will rise over time (with the net price rising exponentially, as before), and output will fall until eventually, at precisely the moment when the first deposit is exhausted, the price will have reached the level at which it pays the second-lowest cost producer to enter the market. The price will continue to rise, and output will continue to fall, until in due course the second source is exhausted, once again at exactly the moment when the next-lowest-cost producer is tempted to start production. And so it goes. Eventually, at just the point at which the most expensive deposit has been exhausted, the price will reach a level sufficient to choke off demand entirely, or alternatively to stimulate the entry of a new resource or new technology capable of substituting for the original resource.

To the skeptical reader this theoretical result will seem suspiciously tidy. And indeed, real-world markets usually aren't nearly as well behaved as the theory suggests. In some cases the resource may be controlled by a single large monopolist or by an oligopoly (such as occurred in the world oil market with the OPEC cartel in the 1970s and 1980s). To maximize the market price of the resource (and hence profit), a monopolist typically restricts output below the rate predicted by the competitive market theory. In other situations competition may result in higher-than-predicted output. For example, when several independent private producers simultaneously have access to the same reservoir of oil or gas, each operator will be motivated to produce quickly, out of the fear that his rivals will get there first and cut off his source of supply. The overall effect is a more rapid exploitation than the theory predicts.

But suppose that neither of these conditions applies, and that the actual production rate closely approximates the predictions of competitive market theory. Does the rate at which resources are exhausted in that case correspond to the socially optimal path? In other words, does even a well-functioning competitive market allocate resources over time correctly? This question does not have a clear answer, not least because of the wide range of views within society as to what level of commitment to conservation is appropriate – views that, as previously noted, may be informed as much by philosophical and ethical considerations as they are by economic reasoning. Nevertheless, our economic model provides insights into this question as well.

It is possible to show that if resources are allocated over time according to the predictions of the competitive market model, the result will be to maximize the present value of the future benefits to society, assuming that society wants to discount these future benefits at the same rate as the market rate of interest, r. So the question of what is the socially optimal rate of resource depletion becomes a question of what is the appropriate social discount rate to use. As we discuss below, many people argue that the social rate of discount is lower than the private market interest rate. According to this view, private producers, making decisions based on the market interest rate, will produce more than they "should." To see why, recall the basic result that producers will always be inclined to produce if they expect to earn a higher return by investing the proceeds from the sale than by holding the resource in the ground. So if production decisions are based on an interest rate that is higher than the social discount rate, the resource will be exploited and therefore exhausted faster than it ought to be.

The correct value of the social discount rate is a very important question, since it determines the intergenerational distribution of the benefits of a fixed resource endowment (or it would do so, if it were imposed on producers). To argue that the social discount rate is lower than the private market rate is to say that private individuals will discount their own future utility and the utility of future generations more heavily than is optimal for society. Many economists and others believe this to be true. They argue that while individuals may worry about their children, and perhaps their children's children, subsequent generations do not figure as strongly in their calculus as would be appropriate from a societal point of view. Some even go so far as to say that there is no reason to treat generations unequally at all – indeed, that it is ethically indefensible to do so – and that this generation, therefore, ought to behave as if the social discount rate were zero. Note that even in this case producers would still produce some of the resource today. But the rate of exhaustion would obviously be lower. (The actual level would depend on assumptions about the rate of technological progress and the substitutability of alternatives.) Others assert that the social discount rate is several percentage points below the market rate.

In theory, if there were broad consensus on the value of the social discount rate, one way to conserve resources would be to require private resource owners to use this rate in their production decisions. But that kind of interference in private decision-making would be unacceptable in market economies. Conservation advocates pursue their objectives in other ways. For example, they seek to prohibit or restrict the development of specific deposits (where a parallel and often primary goal is to protect the environment). Sometimes they seek to tax current production or consumption or to impose restrictions on resource use (such as occurred with natural gas in the 1970s and 1980s.) These measures are controversial, and often conflict not only with the interests of private producers, but also with the societal goal of improving the efficiency of the resource allocation process. (Recall the result that, for any given value of the interest rate, competitive markets will yield a production path through time that maximizes the discounted stream of benefits to society.) The actual process of parcelling out finite resource endowments is thus invariably an untidy one, and whether it adequately recognizes the rights and interests of future generations can certainly be debated. One important and relatively uncontroversial role for governments is to ensure that the latest and best-quality information about the extent of resources and reserves is collected and broadly disseminated. That way, participants in the marketplace – both producers and consumers – can make well-informed decisions, a result that is valuable to everyone.

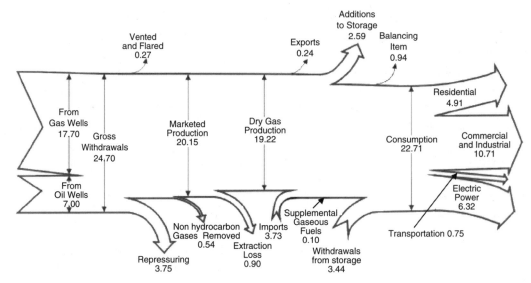

Source: Reprinted from U.S. Energy Information Administration, http://www.eia.doe.gov/emeu/aer/diagrams/diagram3.html.

Figure 10.5. Natural gas flows in the U.S. economy, 2000 (trillion cubic feet).

FUTURE NATURAL GAS USE IN THE U.S. AND THE ECONOMICS OF GAS-FIRED ELECTRICITY GENERATION

The current uses of natural gas in the United States are shown in Figure 10.5. Natural gas use is expected to increase by an average annual rate of 2% between 2000 and 2020. As discussed earlier, this is mainly because most replacement and expansion of electricity generating capacity in the United States over the next twenty years will come from natural gas (see Figure 10.6).

The future cost of electricity from natural gas will depend importantly on the price of natural gas. But a simple analysis shows just how attractive this form of electricity generation is today. First, natural gas combined-cycle plants that utilize both combustion and steam turbines are highly efficient; we will assume an efficiency of 45%. (This can be compared with coal plants that typically operate in the range of 33%.) Second, capital costs are quite low for gas plants compared with either coal or nuclear plants. A combined-cycle plant is likely to cost about $600/kWe capacity. Third, these plants are typically highly reliable and operate with availabilities of 90% or higher. Fourth, operation and maintenance costs are quite low, perhaps 5 mills/kwe-hr. Fuel costs will of course depend on the price of natural gas.

Assuming an interest rate of 10% per year and a plant lifetime of twenty years, the annual capital charge rate is 0.117 per year.[3] If the capacity factor of the plant is 90%,

---

[3] Recall from equation (1) in Chapter 3 that the expression for the annual capital charge rate, $\phi = \frac{r(1+r)^N}{(1+r)^N - 1}$, where r is the interest rate and N is the period for which financing is available (e.g., the loan term.)

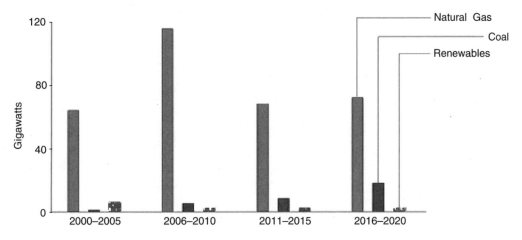

Figure 10.6. Projected electricity capacity additions in gigawatts by fuel type, 2000–2020.

the capital component of the electricity cost (in cents per kilowatt hour) is

$$600(\$/\text{kwe}) \times 0.117(\text{yr}^{-1}) \times 100 \ (\cent/\$) \times \frac{1}{8766} \left(\frac{\text{yrs}}{\text{hr}}\right) \times \frac{1}{0.9} = 0.89\cent \text{ per kwhr.}$$

As of this writing (mid-2002), the natural gas acquisition cost for U.S. utilities is about $3 per thousand cubic ft (MCF). Thus, the fuel cost component of the electricity cost (in cents per kilowatt-hour) is

$$3\left(\frac{\$}{\text{MCF}}\right) \times 10^{-6} \left(\frac{\text{MCF}}{\text{BTU}}\right) \times 3412 \left(\frac{\text{BTU}}{\text{kwhr}}\right) \times \frac{1}{0.45} \left(\frac{\text{kwhr(th)}}{\text{kwhr(e)}}\right) \times 100 \left(\frac{\text{cents}}{\$}\right)$$

$$= 2.34\cent \text{ per kwhr.}$$

Assuming an O&M cost of 0.5¢ per kwhr, the total cost of gas-fired electricity today is about 3.7¢ per kwhr, which is very competitive with most alternative sources of electricity in the United States.

In previous chapters we have considered the generation of electricity from wind, coal, and nuclear power. Table 10.4 compares the production costs of electricity from these different technologies for the reader's convenience. The economic advantage of natural gas under current conditions is evident.

Several other points should also be made about the comparisons in Table 10.4. First, the wind cost estimate does not incorporate the 1.5¢ per kwhr federal renewable energy tax credit, which obviously enhances the competitiveness of wind power. On the other hand, the wind cost estimate also takes no account of the additional cost of backup power. Because of the intermittency of the winds, backup power supplies (or

Table 10.4. Comparison of electricity generating costs

Technology	Capital cost ($/kwe)	Capacity factor (%)	Operating life (years)	Capital charge factor[4] (yr^{-1})	Capital cost (c/kwhe)	O&M cost (c/kwhe)	Fuel cost (c/kwhe)	Total cost (c/kwhe)	Lifetime levelized cost[5] (c/kwhe)
Wind[1]	1,100	32.5	13	0.141	5.44	0.9	0	6.34	6.56
Coal[2]	1,500	80	30	0.106	2.27	0.7	1.5	4.47	5.49
Gas	600	90	20	0.117	0.89	0.5	2.34	3.73	4.79
Nuclear[3]	2,500	80	30	0.106	3.77	0.9	0.65	5.33	6.03

[1] The capital cost and capacity factor estimates for the wind system were obtained from M. Jacobson and G. Masters, "Exploiting wind versus coal," *Science 293*, p. 1438 (August 24, 2001).

[2] The coal plant capital cost, capacity factor, and fuel cost estimates were first presented in Chapter 4.

[3] The nuclear cost estimates appear here for the first time.

[4] Assuming an interest rate of 10%/year.

[5] The levelized costs are calculated over the operating life of each technology. All fuel and O&M costs are assumed to escalate at 4%/year. (Since the operating lifetimes of the technologies are different, these levelized costs are not strictly comparable across technologies.)

energy storage) would in practice be necessary if the wind farm was to function, like the other technologies in the table, as a baseload source of power.

Table 10.4 presents comparisons both of today's production costs and the lifetime levelized production costs. As discussed in Chapter 4, the lifetime levelized cost is a better measure of economic merit because it takes into account changes in fuel and other costs over the plant lifetime (to the degree that these can be accurately forecast.) The expression for the fuel component of the lifetime levelized cost, e_F, is[4]

$$e_F = e_F^o \frac{r}{r - x} \left[\frac{1 - e^{-(r-x)L}}{1 - e^{-rL}} \right],$$

where e_F^o is the fuel cost at the beginning of plant life, x is the expected annual rate of increase in the fuel cost, r is the interest rate and L is the levelization period. The O&M cost is levelized in a similar manner. For simplicity, all annually recurring costs (i.e., fuel and operating and maintenance expenses) are assumed to escalate at the same rate.

APPENDIX TO CHAPTER 10

We closely follow the treatment of R. S. Pindyck.[5] In a competitive market, producers take the price $p(t)$ as given, and choose a rate of production $q(t)$ from a reserve base $R(t)$ so as to maximize the net present value (NPV) of the reservoir. If the cost of production is $c(t)$, we have that

$$\text{NPV} = \int_0^\infty e^{-rt} \{ p(t)q(t) - c(t)q(t) \} dt,$$

where r is the discount rate. In general, the cost of production $c(t)$ will depend on the size of the reserve, $R(t)$, that is, $c(t) = c[R(t)]$. Early in the life of the reservoir, the cost of extraction is low, whereas at later times the cost will be higher. Eventually, this rising cost of extraction will lead to abandonment of the reservoir.

A resource owner wants to choose the production path $q(t)$ that will maximize the NPV of the property given the price trajectory $p(t)$ in the marketplace. But the total production cannot exceed the size of the reservoir R. Thus, there is a constraint at every point in time:

$$q(t) = \frac{dR}{dt}.$$

Thus, we must find the optimal path to maximize the NPV subject to the constraint. This is done by adding a Lagrange multiplier $\lambda(t)$ to the NPV expression

$$\text{NPV} = \int_0^\infty dt \left\{ e^{-rt} [p(t)q(t) - c(t)q(t)] + \lambda(t) \left[q(t) - \frac{dR}{dt} \right] \right\}.$$

From the calculus of variations we know that the effect on NPV of a variation $\delta q(t)$ is

$$\delta\text{NPV} = \int_0^\infty dt \{ e^{-rt} [p(t) - c(t)] + \lambda \} \, \delta q(t).$$

[4] See equation (2) in Chapter 4.
[5] Robert S. Pindyck, *J. of Political Economy*, 86, 840 (1978).

A necessary condition for NPV to be a maximum is that this variation vanishes, that is,

$$e^{-rt}[p(t) - c(t)] = -\lambda(t).$$

The variation of NPV with reservoir size is

$$\delta\text{NPV} = \int_0^\infty dt \{-e^{-rt}c'[R(t)]q(t)\delta R(t) - \lambda(t)\delta\dot{R}(t)\}.$$

Integration by parts on the second term in the integral gives

$$\int_0^\infty dt \{\lambda(t)\delta\dot{R}\} = \lambda(t)\delta R(t)\Big|_0^\infty - \int_0^\infty dt \{\delta R(t)\dot{\lambda}(t)\}.$$

The first term on the right-hand side of this equation vanishes, because the reservoir values are fixed at the beginning and the end of production. Thus,

$$\delta\text{NPV} = \int_0^\infty dt \{-e^{rt}c'[R(t)]q(t) + \dot{\lambda}(t)\}\delta R(t)$$

from which it follows that

$$\dot{\lambda}(t) = e^{-rt}c'[R(t)]q(t).$$

For the simple case where the unit cost of extraction is constant, $c(t) = c[R(t)] = c$ and the Lagrange multiplier is constant in time. Thus,

$$e^{-rt}[p(t) - c] = -\lambda = \text{a constant.}$$

The quantity $(-\lambda)$ is the present worth of a unit of future production. Taking time derivatives, we find an expression for λ

$$\dot{p}(t) = -\lambda r e^{+rt},$$

and substituting for λ we have the equation

$$\dot{p}(t) = r[p(t) - c],$$

which can immediately be integrated to give the result in the text

$$\pi(t) = p(t) - c = \pi(0)e^{rt}.$$

In a competitive market, producers will move in and out of the market until the market clearing price is exactly on this trajectory. The NPV at the market clearing price is

$$\text{NPV} = \int_0^\infty dt \{e^{-rt}[p(t) - c]q(t)\} = \int_0^\infty dt \{[p(0) - c]q(t)\}.$$

This says that there will be either zero production or production at maximum capacity, depending on whether the market price exceeds the unit production cost. This is a thoroughly sensible result; what is new is the expectation in a competitive market of exponential growth in the net price or economic rent.

11

Safety and Risk: Examples from the Liquefied Natural Gas and Nuclear Industries

The application of almost all new technologies involves some degree of risk to public health and safety and the environment, and these risks must be systematically considered. For some technologies the risks are primarily confined to the manufacturing process. In other cases, it is the users who incur the main risks. In still other cases, the risks are externalized – that is, they are borne by people who are not direct beneficiaries of the technology either as suppliers or users. Where the new technology is displacing an existing product or process, the net risk to society may be either increased or reduced. A few technologies have the potential to cause harm on a large scale as a result of a single event. The probability of such events may be extremely low, but they cannot be ruled out entirely. Special methods have been developed to evaluate these low-probability, high-consequence risks. This chapter briefly introduces these methods, using nuclear power plants and liquefied natural gas facilities as examples. We also consider the question of public attitudes toward health and safety risks. Innovators and safety regulators alike need to understand how the public perceives risks, how these perceptions are formed, and what causes them to change.

LIQUEFIED NATURAL GAS

There are many areas of the world where gas exists in great abundance, either in free deposits, for example, in New Zealand, Indonesia, and Algeria, or associated with oil reserves, for example, in Nigeria and Saudi Arabia. In some oil producing regions so much associated gas is produced in the process of oil drilling that it becomes a disposal problem. A certain amount of gas can always be re-injected into the oil field to maintain pressure, but much excess gas is often simply flared. In countries where the gas supply is so plentiful that it exceeds domestic requirements, there is understandable interest in seeking ways to export the gas to overseas markets. In bringing the gas to market there are three basic choices to consider.

The gas can be shipped by pipeline. Russia exports vast quantities of gas from Siberia to Western Europe by pipeline. There is also continuing interest in building a pipeline to carry gas from Alaska to the lower forty-eight states of the United States, as noted in the previous chapter.

The second choice is to convert the gas to a liquid product. Saudi Arabia and other Middle Eastern oil-producing countries have built extensive petrochemical complexes to transform natural gas, as well as petroleum and natural gas liquids such as butane, to higher-value liquid products that are more efficiently shipped to distant

Source: Photo courtesy Atlantic LNG.

Figure 11.1. Atlantic LNG Facility, Point Fortin, Trinidad and Tobago.

markets. Examples of bulk liquid chemicals made from natural gas are methanol and ammonia.

The third choice is to liquefy the natural gas at very low temperatures at port, load the liquefied natural gas (LNG) onto specially constructed tankers, and ship the LNG to market. The LNG must be off-loaded at a terminal facility where it can be gasified and then distributed by conventional pipeline. A large natural gas liquefaction facility is shown in Figure 11.1.

Currently the major exporters of LNG are Algeria and Indonesia, while the major LNG importing nations are France and Japan. The United States is importing increasing amounts of LNG. There are three operating LNG terminals in the United States: Everett, Massachusetts; Cove Point, Maryland; and Lake Charles, Louisiana.

The use of LNG raises several issues. First, there is the political issue of reliance on a foreign supplier. LNG imports present many of the same problems of dependence as do imports of oil.

Second, LNG is a relatively high-cost form of natural gas because of the added costs of liquefaction, tanker transportation, and gasification. Even if the imputed price of gas paid to the producer at the well-head is low or zero, the final cost required to get the gas to market is high. This means that LNG can only compete in end-use markets where the price of alternative fuels is high. This is true in most of Europe where natural gas displaces diesel fuel (No. 2 oil), which has a relatively high price. It is less likely to be true in U.S. markets, where natural gas is used mostly for heating and thus competes with lower cost residual oil (No. 6 oil).

The early LNG contracts between U.S. firms and foreign natural gas suppliers, mostly signed in the mid-1970s, became very controversial. When the price of gas in U.S.

end-user markets rose dramatically during the oil price shocks of 1978–79, Algerian gas suppliers demanded higher well-head prices. Fierce disputes arose between the LNG suppliers and LNG importers over this issue. Since then much shrewder joint venture arrangements have been negotiated which make the gas producer and the LNG importer share the benefits and the risks of the project in both good times and bad. This avoids pitting the gas supplier against the LNG importer.

The third issue presented by LNG is safety. What is the risk that a tanker carrying a full cargo of LNG will explode – in the worst case, in a densely populated harbor? This is the issue we wish to examine in this chapter. What is the probability of such an accident? If there is an accident, to what extent will the public be exposed to damage? What is the consequence of this exposure?

LNG is only one example of the general problem of managing technological risk in our society. Reactor safety, discussed briefly in Chapter 7 and again later in this chapter, is another important example. Many non-energy-related activities also pose significant risks to the public, such as the production and transportation of toxic chemicals, highway travel, and air transportation. In each of these activities there is the possibility that an accident will occur, that the public will be exposed to danger, and that damage will result. In order to manage such risks effectively, several steps must be taken. First, the probabilities of accidents and of subsequent exposures must be estimated, along with the consequences of exposure. Second, effective strategies for risk prevention and mitigation must be developed. And third, since public perceptions of risks may be quite different from what the statistics suggest to a risk specialist, the task of informing the public and responding to its expressed concerns is also an integral part of risk management.

The LNG Accident Scenario. A large LNG tanker carries about 125,000 m^3 of methane (about 60 thousand metric tons) into port. The enormous size of such tankers is apparent in Figure 11.2. The methane is contained in pressurized steel tanks. A serious accident would involve a break in a LNG tank vessel, followed by the dispersal of the liquid methane cargo into a cloud, and then the detonation of the fuel-air mixture by some initiating event. The potential energy release can be estimated from the heat of combustion of the natural gas:

$$CH_4 + 2O_2 \rightarrow CO_2 + 2H_2O \qquad \Delta H = -802 \text{ KJ/mole.}$$

A crude calculation indicates that the explosion would be equivalent to about 20 kilotons of TNT, approximately the size of the atomic bomb dropped on Hiroshima. Evidently such an explosion would cause enormous damage if the accident occurred at a terminal located close to a population center. Of course, any liquid hydrocarbon cargo of the same weight has roughly the same heat of combustion, and an explosion of approximately the same size would be predicted if a credible dispersal and detonation sequence could be postulated. LNG is different from other hydrocarbons because in this case dispersal into a fuel-air cloud that could be detonated, although very unlikely, is at least plausible.

Source: Photo courtesy of Tractebel LNG North America.

Figure 11.2. A Liquefied Natural Gas Tanker in Boston Harbor.

Probabilistic Risk Assessment

An analytic technique has been developed to estimate systematically the risk of accidents of this type. The technique, called probabilistic risk assessment (PRA), seeks systematically to address three questions: (1) What can go wrong? (2) How likely is it that this will occur? and (3) What will be the outcome? The method was originally used in the U.S. space program, and was further developed by a team led by Professor Norman Rasmussen of MIT in order to address the problem of nuclear reactor safety. Its first major application to nuclear power plants occurred in the famous Reactor Safety Study published by the Nuclear Regulatory Commission in 1975, also known as the WASH 1400 report, and the method has since come to be widely used in the nuclear industry.[1] PRA has also been applied to other technologies, including chemical plants, airplanes, and LNG facilities.

The method starts from a definition of risk as the consequences of the activity (i.e., the number of fatalities, or injuries, or the cost of physical damage) that are expected

---

[1] For a review, see Norman C. Rasmussen, "The application of probabilistic risk assessment techniques to energy technologies," *Ann. Rev. of Energy*, 6, 123–138 (1981).

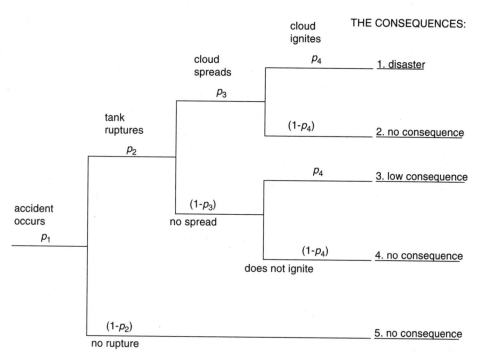

Figure 11.3. LNG tanker accident scenario.

to occur per unit time. The risk can then be written as:

$$\text{risk (consequences/unit time)} = \text{frequency (event/unit time)}$$
$$\times \text{magnitude (consequence/event)}.$$

To evaluate the risk the PRA method uses two key tools: event tree analysis and fault tree analysis. The first step in the event tree analysis is to select an initiating event. Inductive reasoning is then used to identify all possible outcomes of that event, including those that might lead to an accident. Probabilities are then estimated for each event on the pathway to the accident.

The probability of an undesired event can be estimated using fault tree analysis. The fault tree shows the combinations of faults that can lead to the failure of a system or a sub-system. The top level of a fault tree is the undesired event. The faults which lead to it are the branches descending from the top. At the bottom of the tree are the primary events that contribute to the system failure. Empirical data are used to estimate the probability of these primary events, and algebra to obtain the combined probabilities.

A highly simplified example of the application of PRA to LNG safety is presented in Figure 11.3. The simplified event tree description of an accident involving an LNG tanker is shown in the Figure. Note that the sum of the probabilities of all outcomes equals p_1, the probability of the initiating event. The probability of a disaster (outcome #1) is simply equal to $p_1 \times p_2 \times p_3 \times p_4$. The probability of no consequence given an accident is equal to the sum of the probabilities of outcomes $2 + 4 + 5 = p_2 \times$

$(1 - p_4) + (1 - p_2)$. The probability of a small consequence, given an accident, is the probability of outcome $3 = p_2 \times (1 - p_3) \times p_4$. Once an estimate is made of these various probabilities, the likelihood of a severe accident can be assessed.

The PRA method has significant strengths. First, it forces systematic attention to accident scenarios. Second, it helps to structure and clarify debates over assumptions about accident scenarios and the likelihood of their occurrence. Third, analysis of the event trees and fault trees can identify events and systems that are especially important to outcomes and that therefore deserve engineering attention. Finally, the PRA method provides a much-needed common language for the regulatory process.

But the PRA method also has some serious weaknesses. First, the definition of the event tree is not verifiable. There is always the question of whether all of the most important initiating events have been identified. Another problem is that there is often no objective basis for the estimates of event probabilities. This is because empirical data are often lacking for the key events and there is no rigorous way to estimate the probabilities.

A particularly important limitation of the PRA method is that the list of events between initiation and consequence is not bounded. If the events are considered to be independent, then adding events to the chain will inevitably lead to small probabilities, because:

$$\lim_{n \to \infty} \prod_n p_n \to 0.$$

Of course, the events may not be independent, and if they are not a different problem arises. This is the "common mode" failure problem: How does one deal with the possibility that the events in the postulated accident sequence are related by a common failure mode that has not been identified? If a common failure mode exists, then the small probability estimated by treating the sequence of events as independent must be replaced by the single probability of occurrence of the common mode. Postulating the existence of an unidentified common failure mode evidently implies a higher estimate for the probability of an accident.

RISK PERCEPTIONS

A careful, thorough PRA may be very convincing to the trained specialist. But attitudes towards risk among the general public are often quite different from the specialists' understandings. People often express great anxiety about hazards that technical analyses indicate pose very low risks, yet are indifferent to other hazards about which experts are much more concerned. The experts, as noted above, measure risk as the product of probability and consequence. But public risk perceptions seem to be influenced by other factors, too. For example, the technical measure of risk does not differentiate between activities that have a high likelihood of causing a small number of fatalities and those that have a low likelihood of causing a large number of fatalities. If the expected number of fatalities is the same, the risk, according to this measure, is also the same.

Yet many people seem to be much more concerned about low-probability accidents with high consequences. This leads us to consider two fundamental questions: How are people's perceptions and beliefs about risks formed? What causes these perceptions to change?

One study conducted by psychology researchers asked members of the public to rank order a diverse group of 30 hazards according to the risk of death they posed.[2] Three separate groups of laypeople were surveyed in the study, along with a fourth group comprised of specialists in risk assessment. The results are summarized in Table 11.1. There are some striking differences between the lay groups and the experts. Nuclear power was rated as the riskiest of all the activities by two of the lay groups and the eighth riskiest by the third, but was ranked only twentieth by the experts.

The risk judgments of the experts were closely correlated with actual or estimated technical statistics on annual fatalities. (For hazards such as handguns and highways it is possible to count the victims, but in other cases – such as nuclear power – fatality estimates must be made based on inference.) But the risk judgments of the laypeople were not strongly related to these statistics. There are two possible explanations for this divergence. One is that members of the public base their judgments about risks on factors other than expectations of annual fatalities. The other possibility is that public risk perceptions actually are based on expectations of fatalities, but that these expectations are inaccurate. To test the latter possibility, the researchers asked the laypeople to estimate how many people are likely to die in a typical year from each of the 30 activities and technologies. These subjective fatality estimates are shown in Table 11.2 together with the technical fatality statistics. It turns out that neither set of estimates is a good predictor of public risk perceptions. The laypeople's risk rankings are no more closely correlated with their own subjective estimates of fatalities than they are with the actual fatality statistics. Clearly, then, people do not equate risk with annual fatalities. Other factors must be at work in shaping public risk perceptions.

There was a particularly glaring discrepancy between the public's perception of nuclear power risks, which as noted previously were ranked highest of all by two of the lay groups, and the same two groups' subjective estimates of annual fatalities from nuclear power, which were the lowest of all. What could account for such a difference? One possible explanation is that people have a special fear of nuclear radiation. Yet the risks from another source of radiation, medical X-rays, were ranked much lower by the laypeople – lower, in fact, than the experts ranked them (see Table 11.1). Fear of radiation per se does not seem to be the root cause of the discrepancy.

Another possible explanation is that the laypeople ranked nuclear power risks as high as they did because of this technology's perceived potential for disaster. To test this possibility, the researchers asked their subjects what they expected the fatality count to be in a particularly bad year from each type of hazard. The results are shown in the

[2] Paul Slovic, Baruch Fischhoff and Sarah Lichtenstein, "Facts and Fears: Understanding Perceived Risk," in R. C. Schwing and W. A. Albers (eds), *Societal Risk Assessment: How Safe is Safe Enough?* (Plenum, New York, 1980), p. 181–216.

Table 11.1. Ranking of perceived risks from 30 activities and technologies (in descending order of perceived risk)

	Group 1 League of Women Voters	Group 2 College students	Group 3 Active club members	Group 4 Risk experts
Nuclear power	1	1	8	20
Motor vehicles	2	5	3	1
Handguns	3	2	1	4
Smoking	4	3	4	2
Motorcycles	5	6	2	6
Alcoholic beverages	6	7	5	3
General (private) aviation	7	15	11	12
Police work	8	8	7	17
Pesticides	9	4	15	8
Surgery	10	11	9	5
Fire fighting	11	10	6	18
Large construction	12	14	13	13
Hunting	13	18	10	23
Spray cans	14	13	23	26
Mountain climbing	15	22	12	29
Bicycles	16	24	14	15
Commercial aviation	17	16	18	16
Electric power	18	19	19	9
Swimming	19	30	17	10
Contraceptives	20	9	22	11
Skiing	21	25	16	30
X-rays	22	17	24	7
High school & college football	23	26	21	27
Railroads	24	23	20	19
Food preservatives	25	12	28	14
Food coloring	26	20	30	21
Power mowers	27	28	25	28
Prescription antibiotics	28	21	26	24
Home appliances	29	27	27	22
Vaccinations	30	29	29	25

Source: Paul Slovic, Baruch Fischoff, and Sarah Lichtenstein, "Facts and Fears: Understanding Perceived Risk," in R. C. Schwing and W. A. Albers (eds.), *Societal Risk Assessment, How Safe is Safe Enough?*, Plenum, New York, 1980, p. 181–216.

two right-hand columns of Table 11.2. Evidently the lay groups thought that nuclear power had a far greater potential for disaster than the other activities. This result quite clearly suggests that the public's view of nuclear risks is indeed shaped by its fears of a catastrophic nuclear accident.

Further research has pointed to the influence of several other qualitative risk factors, or "attributes," on public risk perceptions. These include:

- Controllability   To what degree can people exposed to the risk avoid death by their own skill or diligence?

Table 11.2. Fatality estimates and disaster multipliers for 30 activities and technologies

		Geometric mean fatality estimate (average year)		Geometric mean multiplier (disastrous year)	
	Technical fatality estimates	League of Women Voters	College students	League of Women Voters	College students
1 Smoking	150,000	6,900	2,400	1.9	2.0
2 Alcoholic beverages	100,000	12,000	2,600	1.9	1.4
3 Motor vehicles	50,000	28,000	10,500	1.6	1.8
4 Handguns	17,000	3,000	1,900	2.6	2.0
5 Electric power	14,000	660	500	1.9	2.4
6 Motorcycles	3,000	1,600	1,600	1.8	1.6
7 Swimming	3,000	930	370	1.6	1.7
8 Surgery	2,800	2,500	900	1.5	1.6
9 X-rays	2,300	90	40	2.7	1.6
10 Railroads	1,950	190	210	3.2	1.6
11 General (private) aviation	1,300	550	650	2.8	2.0
12 Large construction	1,000	400	370	2.1	1.4
13 Bicycles	1,000	910	420	1.8	1.4
14 Hunting	800	380	410	1.8	1.7
15 Home appliances	200	200	240	1.6	1.3
16 Fire fighting	195	220	390	2.3	2.2
17 Police work	160	460	390	2.1	1.9
18 Contraceptives	150	180	120	2.1	1.4
19 Commercial aviation	130	280	650	3.0	1.8
20 Nuclear power	100[a]	20	27	107.1	87.6
21 Mountain climbing	30	50	70	1.9	1.4
22 Power mowers	24	40	33	1.6	1.3
23 High school & college football	23	39	40	1.9	1.4
24 Skiing	18	55	72	1.9	1.6
25 Vaccinations	10	65	52	2.1	1.6
26 Food coloring	_[b]	38	33	3.5	1.4
27 Food preservatives	_[b]	61	63	3.9	1.7
28 Pesticides	_[b]	140	84	9.3	2.4
29 Prescription antibiotics	_[b]	160	290	2.3	1.6
30 Spray cans	_[b]	56	38	3.7	2.4

[a] Authors' estimates based on statistical inference.
[b] No estimate reported.

Source: Paul Slovic, Baruch Fischoff, and Sarah Lichtenstein, "Facts and Fears: Understanding Perceived Risk," in R. C. Schwing and W. A. Albers (eds.), *Societal Risk Assessment: How Safe is Safe Enough?* (Plenum, New York, 1980), p. 181–216.

- Immediacy of effect — Is the risk of death immediate, or more likely to occur at a later time?
- Severity of consequences — How likely is it that the consequence of an accident will be fatal?

- Knowledge about the risk   To what extent is the nature of the risk understood by
                                    those exposed to it, and by the scientific community?
- Dread   Is the risk one that people have learned to live with,
                                    or is it one that inspires feelings of dread?

Many scientist and engineers, frustrated by what they see as irrational fears about technological risks among the general public, blame media bias and sensationalism. They urge less sensational and more balanced reporting, as well as more intensive education efforts to inform the public about actual mortality rates associated with different technologies. Sensationalism and bias in the media surely do affect risk perceptions negatively, and raising the level of public education about risk is surely a very desirable goal for any democratic society. But studies such as the one described here suggest that what is often dismissed as irrational by technical professionals may in fact have deeper roots; public risk perceptions depend in particular and predictable ways on certain qualitative characteristics of hazards and there is a need for a broader definition of risk that encompasses these characteristics, as well as quantitative measures of impact. Some researchers have concluded that such characteristics are actually more important in determining risk perceptions than quantitative measures like expected annual mortality that are usually preferred by scientists and engineers.[3]

What conclusions should technical practitioners and policymakers draw from such findings? One approach would be to discount the significance of qualitative risk attributes like "dread" and "controllability" and continue to insist on the primacy of quantitative measures, such as expected mortality, in making design and technology selection decisions, on the grounds that this is the only rational basis on which to allocate society's resources. The fact that public perceptions are at least partly based on qualitative characteristics is, from this perspective, an aberration. The problem can be ameliorated by an energetic risk communication strategy that stresses the importance of quantitative risk measures and seeks to bring public risk perceptions more closely into line with them.

A quite different approach would be to accept that the quantitative risk measures preferred by scientists and engineers do not conform to society's risk preferences, and that technical choices should take account of the full range of influences on these perceptions.

Each of these approaches presents problems. If technical professionals ignore the factors influencing public perceptions of risk, their efforts to innovate are more likely to be rejected in the court of public opinion. If, on the other hand, they give extra weight to qualitative factors such as "dread" and "potential for catastrophe," the cost of achieving a given level of safety (as measured by avoided loss of life) may be greater. Indeed, it is

---

[3] C. Hohenemser, R. W. Kates, and P. Slovic, "The Nature of Technological Hazard," *Science*, 220, 378–384, (1983).

possible that the actual benefit in terms of avoided loss of life would be smaller than if design decisions were dictated by quantitative criteria.

An example of the interplay of these issues has arisen in the current debate about the future of nuclear power. One of the proposed strategies for addressing public concerns over nuclear plant safety is to introduce "inherently safe" nuclear reactors. It will be recalled from Chapter 7 that the safety of the current generation of light water power reactors is based on a "defense-in-depth" philosophy. A complex array of active cooling systems ensures with very high probability that in any accident situation the radioactive decay heat will be removed from the core before the fuel overheats. Because the metal oxide fuel rods would melt in a matter of minutes if cooling were lost, much redundancy is built into the design, including backup cooling systems, backup power supplies, and conservatively designed pumps and piping. The possibility that all these safety systems will fail at the same time cannot be completely ruled out, but probabilistic risk assessments predict that the probability of core melt in a large, modern PWR is less than 1 in 10,000 per year.

An alternative approach to safety is to design the reactor such that, even in the event of a complete loss of coolant, natural heat removal processes would suffice to remove the decay heat. In this case there would be no need to rely on the proper functioning of backup cooling systems and the correct intervention of reactor operators to achieve emergency cooling. This is the principle of *passive safety*. A reactor concept that embodies this principle is the helium-cooled, graphite-moderated pebble-bed reactor shown in Figure 11.4. The core of this reactor is comprised of hundreds of thousands of tennis-ball-sized graphite "pebbles," each of which contains thousands of tiny uranium oxide particles half a millimeter in diameter, coated with multiple layers of carbon and silicon carbide to prevent the escape of fission products. The core is normally cooled by pressurized helium gas at high temperature. In some versions of the concept, the helium is piped directly to a gas turbine. In others, the hot gas is used to raise steam which is then sent to drive a steam turbine.

The helium gas has a very low heat capacity, so if the flow of coolant through the core is interrupted essentially all of the decay heat (which of course continues to be generated even after the fission chain reaction has been terminated) is initially absorbed by the graphite pebbles. The pebbles are thermally stable, however, and retain their integrity even at very high temperatures. Even in the worst case scenario, involving withdrawal of the reactor control rods, depressurization of the core and total loss of coolant, the temperature of the core does not rise to dangerous levels from the perspective of fuel stability. In this scenario the combination of natural heat conduction and thermal radiation suffices to remove the decay heat and stops the graphite pebbles from overheating and releasing radioactive fission products.

Proponents of the pebble-bed reactor argue that this 'walkaway safe' system has several intrinsic advantages over the current generation of light water reactors. They point to the capital cost savings stemming from the absence of need for costly emergency core cooling systems and potentially also for the massive concrete containment structure

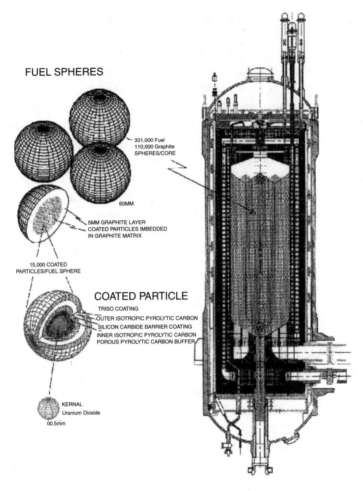

FUEL SPHERES

331,000 Fuel
110,000 Graphite
SPHERES/CORE

60MM

5MM GRAPHITE LAYER
COATED PARTICLES IMBEDDED
IN GRAPHITE MATRIX

15,000 COATED
PARTICLES/FUEL SPHERE

COATED PARTICLE

TRISO COATING
OUTER ISOTROPIC PYROLYTIC CARBON
SILICON CARBIDE BARRIER COATING
INNER ISOTROPIC PYROLYTIC CARBON
POROUS PYROLYTIC CARBON BUFFER

KERNAL
Uranium Dioxide
00.5mm

Source: PBMR (Pty) Ltd, South Africa (2001).

Figure 11.4. Schematic of Modular Gas-Cooled, Pebble-Bed Reactor.

that serves as a barrier of last resort for light water reactors. They also argue that the passive safety approach will make the reactor more acceptable to a public whose aversion to catastrophic events has not been dispelled by the defense-in-depth approach pursued in light water reactors. In this view, actually being able to demonstrate the safe shutdown of a reactor in worst case conditions will be more persuasive to an anxious, skeptical public than having to rely on complex computer simulations and PRAs, as is necessary for light water reactors. The risk perception studies described in this chapter lend support to this view. But how much this more transparent safety strategy would actually be worth remains to be seen. Will it change public perceptions of the potential for nuclear reactor disaster? Will the threat of a terrorist attack with hijacked planes or truckloads of explosives – now so much more plausible following the attacks on New York and Washington on September 11, 2001 – displace the concern over reactor accidents in the pantheon of public fears? Will such a threat necessitate a massive concrete containment structure around such reactors anyway, even though a worst-case loss of coolant accident would not require this? The important point to emphasize is that for nuclear engineers these are not questions that can be ignored or left to others to answer. They are integral to the process of designing nuclear power plants. And they demand a sophisticated understanding not only of technical issues but also of public risk perceptions. In this case, as in many others, the problem of risk cannot be reduced to a mathematical abstraction.

12

Synthetic Fuels

For the foreseeable future the world will rely on oil and gas to meet much of its energy requirements, especially for transportation. Many countries will need to import much, and in some cases, all of the oil they consume. This dependence on oil and on oil imports prompts an interest in exploring technologies that can produce gas and liquid fuels from more plentiful and accessible raw materials.

Conventional liquid petroleum products such as gasoline, diesel fuel, and kerosene are easily obtained by upgrading crude oil in petroleum refineries. Synthetic fuels (often also referred to as synfuels) are oil and gas substitutes that are produced from more plentiful hydrocarbon resources by complex chemical processing. The raw materials for synthetic fuels are tar sands, shale, and coal. The cost of producing synthetic fuels from these resources defines a "shadow price" for oil and gas products obtained from conventional oil and gas resources. If the price of fuel from conventional sources were to rise above the cost of producing synthetic fuels, the market would be expected to switch to producing synthetic fuels in quantity. The price increase might occur either because of cartel action by oil exporting countries, or because of the progressive depletion of oil and gas resources.

This chapter introduces the technical aspects of some of the principal synthetic fuel production processes. The cost of producing synthetic fuels is also discussed. A principal subject of the chapter is to describe the history of the synthetic fuels initiative launched by the United States in the late 1970s. The U.S. synfuels program was a response to the sharp increase in oil prices that resulted from the OPEC-engineered oil market disruptions of that decade. This massive, several-billion-dollar program failed, and it provides an excellent case study of the political difficulties and dangers encountered in mounting large-scale technology development programs in this country.

TECHNICAL ASPECTS OF SYNTHETIC FUELS

The basic problem of producing synthetic liquids or gas from plentiful resources such as tar sands, coal or shale is that these resources are deficient in hydrogen relative to the hydrogen content of the desired products. Thus the central technical challenge is: (1) to produce hydrogen, and (2) to add the hydrogen to the resource. Synthetic fuel technologies differ in how they accomplish these two tasks.

The only economical way to obtain hydrogen is to extract it from water. This is accomplished by the "water-shift" reaction in which carbon monoxide reacts with

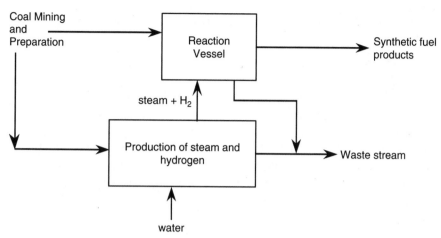

Figure 12.1. Flow diagram for a synthetic fuels process. Coal (or shale or tar sands) is used to produce hydrogen, which in turn is added to the reaction vessel to make the desired liquid or gaseous hydrocarbon product.

steam to produce hydrogen and carbon dioxide:

$$CO + H_2O \rightarrow H_2 + CO_2.$$

The carbon monoxide is made by gasifying the carbon in coal:

$$C + \frac{1}{2}O_2 \rightarrow CO.$$

The CO and H_2 are combined into a product stream referred to as *synthesis gas*, in proportions needed to produce the particular synthetic gas or liquid of interest.

The next step is to add hydrogen to the carbon in the raw materials.

$$carbon + hydrogen \rightarrow liquid\ or\ gaseous\ hydrocarbon\ products.$$

Synthetic fuel technologies differ in how this is done, but the net effect is that the hydrogen-deficient raw material is upgraded to the higher H/C ratio characteristic of liquid or gaseous hydrocarbon products. For each of these processes, complex chemical transformations are carried out and sophisticated chemical engineering is required to make the process work and work efficiently.

The basic flow diagram for synthetic fuels production is illustrated in Figure 12.1.

The cost of the reaction vessel and its operation are a small part of the overall system cost. But the reaction vessel is crucial because it defines the chemical transformations that will take place, the product slate, and the operating conditions and flows required by every other element in the plant. The reaction vessel is where the capital and operating costs of the entire plant are determined. The nature of the reactions and how efficiently they are carried out also determine all the waste products from the synthetic fuel plant and thus its environmental effects.

Three equally important questions must be asked about alternative synthetic fuel technologies:

1. Which of the technologies works best?
2. What will be the relative costs of the alternative technologies when they operate at scale?
3. What are the environmental impacts of the alternative technologies?

The range of alternative technologies is discussed next. There has been some practical experience with each of these technologies. Some of the leading projects of the past two decades (several of which never got beyond the planning stage) are listed below:

Oil from Tar Sands

- 120,000 barrel per day syncrude project in Fort McMurray, Alberta;
- ALSANDS project at Athabasca, Alberta;
- Esso Canada project at Cold Lake, Alberta.

Oil from Shale

- Occidental in-situ retorting at Logas Wash, Colorado;
- Union Oil, surface retorting in Colorado;
- ARCO and Tosco Colony project, shale mining in Colorado.

Synthetic Gas from Coal

- American Natural Resources Great Plains high-BTU coal gasification project in Mercer County, North Dakota;
- Illinois Coal Gasification group in Perry County, Illinois;
- Memphis, Tennessee Light, Gas, and Water medium-BTU project.

Synthetic Liquids from Coal

- Indirect liquefaction: SASOL project in South Africa.
- Direct liquefaction: Ashland Oil's H-Coal project at Catlettsburg; Exxon Donor Solvent project in Texas; Air Products/Wheelabrator-Frye Solvent Refined Coal project SRC I, in Kentucky; Solvent Refined Coal SRC II project in West Virginia.

Syngas. Methane is produced by reacting CO and H_2 over a suitable catalyst according to the reaction:

$$2CO + 2H_2 \rightarrow CH_4 + CO_2.$$

The overall reaction is:

$$4C + 2O_2 + 2H_2O \rightarrow CH_4 + 3CO_2.$$

Note that three molecules of CO_2 are produced for every methane molecule. Hence, burning methane obtained from coal will produce four times as much CO_2 as burning conventional methane. This is one example of the greater environmental impact of synthetic fuels. The syngas process also produces considerable solid waste.

It is possible to write balanced chemical reactions to produce methane which yield fewer molecules of CO_2 per methane molecule. For example, consider

$$2C + 2H_2O \rightarrow CH_4 + CO_2.$$

This is a more favorable reaction from an environmental perspective, but it is not practical at industrial scale. There are many reasons why reactions that can occur in principle are not of practical use: the reaction may be endothermic and require a great deal of energy; the equilibrium may be unfavorable at practical pressure and temperature operating conditions; or the elementary reaction steps may not occur at a fast enough rate and catalysts may not be available to speed them up. Tremendous effort has been expended over the decades by chemical engineers to find chemical reactions, catalysts, and processes that can accomplish these desired transformations efficiently.

When pure oxygen is used to make the syngas, the resulting product is similar in composition to ordinary pipeline quality gas. This product is referred to as "high-BTU" gas. If air is used instead of pure oxygen, then the resulting product will be a mixture of methane and nitrogen (with some environmentally unwanted NO_x). This product is called "medium-BTU" gas because the product (on a volume basis) has lower energy content.

In the past, some industrial plants and cities produced a low-BTU gas that consisted simply of CO produced by burning coal with air. This low-BTU gas, sometimes referred to as "city gas," was used for district heating and gas lighting.

The differences in energy content of these different types of syngas are considerable:

	Energy content (BTU per cubic ft)
High-BTU gas	1,000
Medium-BTU gas	600
Low-BTU gas	300

Coal gasification. It is evident from the preceding discussion that gasifying the coal is crucial to the process. Many different methods of gasification are possible. Traditional gasifiers use fixed, entrained, or fluidized beds. Each has different operating characteristics and results in different waste streams. As mentioned earlier, the operating characteristics of the gasifier determine the operating characteristics of the entire process and hence the cost, as well as the waste streams. Each type of gasifier operates at a

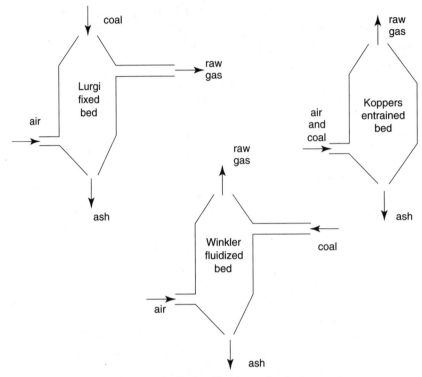

Figure 12.2. Coal gasification technologies.

different temperature and pressure and with a different residence time for the coal and gas stream.

Gasifier	Temperature (°C)	Residence time	
Fixed bed	700–850	coal: 1–5 hr	gas: 15 sec
Fluidized bed	850–1,000	coal: 20–50 sec	gas: 3–5 sec
Entrained bed	1,400–1,500	coal: 1–3 sec	gas: same

A change in the feed coal characteristic requires a change in the operating conditions of the gasifier. Because the gasification reaction is quite exothermic, considerable heat is released. Efficient operation therefore requires attention to heat recovery. The three basic gasifier types are illustrated in Figure 12.2, which gives an impression of the broad range of technical alternatives that exist for gasifying coal.

Coal Liquefaction. Liquids can be produced from coal either directly or indirectly. Indirect liquefaction proceeds from synthesis gas with an appropriate mixture of CO and H_2. The simplest example is the production of liquid methanol, which takes place by passing synthesis gas over a zinc catalyst:

$$CO + 2H_2 \xrightarrow{\text{Zn}} CH_3OH.$$

The indirect liquefaction process was originally developed to convert coal to synthetic liquids. Today, however, methanol is more commonly made starting not with coal but rather with natural gas, which exists in abundance with no local market in many areas of the world such as West Africa. Given high gas prices elsewhere, gas-to-liquid conversion via indirect liquefaction is economically attractive.

More generally the conversion of synthesis gas into a diverse liquid product slate can be achieved with the Fischer–Tropsch process, which uses an iron catalyst. This method was developed by Germany when it was cut off from conventional sources of oil during World War II. The Germans produced essentially all the oil needed by their wartime economy by the Fischer–Tropsch method – a total of about 20,000 barrels per day. (It is interesting to compare this figure with current U.S. oil consumption of 20 million barrels per day.)

South Africa's SASOL applied the Fischer–Tropsch process to produce synthetic liquids from coal during the apartheid era, when the country feared that it would be cut off from all oil supplies. The South Africans built a coal gasification plant based on a Lurgi fixed-bed gasifier that produced one million standard cubic feet of CO per day. The Fischer-Tropsch unit operated at 320 psig and 635°F with a 2.8:1 H_2/CO ratio, and achieved 88% conversion to liquid hydrocarbons. Today SASOL is aggressively marketing this process around the world to convert excess natural gas to liquids.

The indirect liquefaction method has the considerable advantage that all nitrogen and sulfur in the coal is removed in the synthesis gas production step. The disadvantages of this method are that it is thermodynamically inefficient, it produces a diverse product slate, and it requires a great deal of hydrogen.

The other way to produce liquids from coal is by direct liquefaction. This approach adds hydrogen to coal under conditions of high temperature and pressure. The advantage is that the process is thermodynamically efficient and requires less hydrogen per unit of product. The disadvantage is that the nitrogen and sulfur hetero-atoms that are present in the coal remain in the liquid product, making it carcinogenic.

The hydrogen is added to the coal via an organic solvent, a polycyclic aromatic, which is recycled as shown:

No commercial-scale direct liquefaction plants exist today. The DOE sponsored a number of demonstration-scale plants during the 1970s and 1980s: the Solvent Refined

Coal projects in Kentucky (Wheelebrator Frye/Air Products) and in West Virginia (Gulf Oil), and the Exxon Donor Solvent process in Texas.

THE SYNTHETIC FUEL STORY

In the late 1970s the U.S. government adopted a massive program to encourage the production of synthetic oil and gas from coal and shale. The story of this venture resembles a Greek tragedy: the several characters – from the executive branch, the Congress, and industry – each played roles dictated by their inherent characteristics and history that inevitably led to failure. The story is instructive about the forces that shape a massive government effort and the limitations of government intervention which, despite the best intentions, often leads to a waste of public resources and failure.

The Carter administration took office in 1977 against a background of steep increases in energy prices and lines at gasoline stations throughout the nation – the result of OPEC-induced disruptions to the world oil market. Leaders in the administration, notably James R. Schlesinger, the first secretary of the newly created Department of Energy, believed that a strong government response was required to deal with the security consequences of dependence on imported oil and the prospect of further sharp increases in energy prices. Knowledgeable observers realized that the primary response had to be the deregulation of oil and gas prices (in order to moderate demand) and the encouragement of energy efficiency. But there was also a view that beyond these "demand-side" measures there was a need to stimulate new supply options. Many candidate supply options were offered that we discuss elsewhere in this book, including solar energy, wind and gasohol. Still, there was a strong belief that the most important supply initiative should be synthetic fuels, so as to demonstrate the capability of the industrialized world to produce an alternative to conventional oil because of the dependence on unstable OPEC sources of supply.

Several forces prompted the DOE to launch a synthetic fuels program. Energy experts both in and out of government urged the adoption of energy supply initiatives. Congress, reacting to the high energy prices and lines at the gas pump, wanted some aggressive government action to show the people back home that something dramatic was being done. Industry wanted government support to explore technologies that might prove profitable if conventional oil prices continued to rise. Most important, a large amount of money was available because Congress was not prepared to allow the oil companies to reap a "windfall profit" once oil and gas prices were deregulated and domestic energy prices jumped to the higher world levels. It therefore acted to tax these profits away.[1] Thus all the forces needed for government action were present: good intentions on the

---

[1] At that time the federal government regulated the price of natural gas that was transported in interstate pipelines. Oil prices were regulated by controlling the prices that refiners could pay for domestic crude; this low priced oil was blended with high priced imported oil. With deregulation, gas prices could be expected to jump to parity based on the energy equivalence of diesel fuel (No. 2 heating oil) because these fuels were substitutes in industrial boilers. Oil prices would jump because domestic oil would rise to world price levels. Domestic oil and gas producers would thus reap a so-called "windfall profit."

part of government, Congress wanting to respond to citizen concerns, the self-interest of industry (and labor) in seeking government support for new projects, and above all the availability of money.

The purpose of the government program was to help commercialize synfuels. In the United States the production of energy is the responsibility of the private sector. Industry assumes the financial risks of making investments and operating facilities in the expectation of earning an acceptable return on its capital. The massive amounts of capital required come from private sources, not from government expenditures. So there was an important question about what actions it was appropriate for the government to take. Several different types of involvement were possible.

- The government could sponsor R&D on various aspects of synfuels production and make the resulting information available to industry. Everyone accepted this as a legitimate government function, and a robust R&D program, authorized under the Non-Nuclear Energy Act (NNEA), was duly launched.
- The government could assist industry to construct "lead" plants for various synfuels technologies, thus demonstrating the technical characteristics, economic costs, and environmental impacts of the key technologies. This would put the private sector in a better position to expand synfuel production rapidly should market conditions warrant it, because important information needed for private investment would be available.
- The government could stimulate synfuels production capacity by indirect incentives such as tax credits, loan guarantees, equity investments, or guaranteed purchases of product at advantageous prices.
- The government could directly own and operate synfuels plants. Almost no-one believed this was a good idea, since there was no reason to think that the federal government would be as efficient as the private sector at any type of energy production.

The program advanced by the administration had two phases. In the first phase, the government would use indirect incentives to encourage the construction of a few selected technologies. This was expected to cost about $12 billion. In the second phase, indirect incentives would be used to stimulate expanded production to a level of two million barrels of oil equivalent per day. The second phase never happened; had it done so it could have cost as much as $60 billion.

The government chose a novel way to run the synfuels program. Rather than relying on DOE, the Synfuels Corporation (SFC), a quasi-public corporation, was established to manage the effort. Three reasons were advanced for adopting this unconventional approach. First, Congress and the public did not have much confidence that DOE or any other government agency could run large projects efficiently. Creating a quasi-independent, not-for-profit corporation would give greater confidence that the program would be run in a business-like way, in terms of both the objective selection of specific projects and the efficiency of operation. Second, the quasi-corporate structure of the SFC would help achieve the objective of the demonstration phase: to have an operation that would serve as a credible model for possible future private industry investment. The SFC was supposed to function along the lines of a private investment banker. Finally, once the SFC was established and financed (with a one-time appropriation from the accompanying windfall profits tax revenues), there would be less involvement by Congress in the annual authorization and appropriation process. The SFC structure would avoid the federal procurement regulations, personnel rules, and sunshine provisions that constrain public programs.

The authorizing legislation was presented to Congress in 1980 as part of an omnibus piece of legislation called the Energy Security Act of 1980. The SFC, originally called the Energy Security Corporation (ESC), was linked with the windfall profits tax and an ambitious proposal to establish a new Energy Mobilization Board. The purpose of the Energy Mobilization Board was to speed the process required to gain regulatory approval of the construction of major energy facilities. The Congress did not approve establishment of the Energy Mobilization Board; there was no immediate benefit and the measure was passionately opposed by the environmental community. The environmental community also opposed creation of the SFC, but here there was immediate benefit: the expenditure of public funds for projects that created jobs.

The actions of Congress in passing the Energy Security Act are revealing. $20 billion was originally requested for the first phase of the synfuels program. The funds were reprogrammed by Congress as follows:

- $1 billion for solar and conservation R&D;
- $2.5 billion for additional synfuels projects managed by DOE under the NNEA authority;
- $3.0 billion authorized under the Defense Production Act for DOD to purchase synfuel products;
- $1.3 billion for alcohol fuels;
- $12.2 billion remaining for the SFC synfuels program.

In other words, the available money was spread around so there was something for everyone. The renewable energy special interests were taken care of. Congress, not fully trusting the new quasi-independent agency to do its bidding, kept the option open to pursue its own interests in synfuels by conventional mechanisms that it controlled, namely the annual budget process under the NNEA and the Defense Production Act.

The SFC opened its doors and almost immediately came under fire. Congress and the press objected to the salaries that were paid to SFC officers, the fact that meetings were not open to the public, and the method of soliciting projects, which was regarded as favoring large firms. Sharp criticism of the SFC also came from the environmental community, which believed that reducing energy use was the least costly way of reducing dependence on imported oil. Most economists objected to government subsidies of the energy sector.[2] The conventional economic view was (and remains) that energy should

---

[2] Three MIT economics professors, Paul Joskow, Robert Pindyk, and Richard Schmalensee, were especially sharp critics of the SFC.

not be treated differently from any other commodity; once prices were deregulated and free to rise to the level determined by the market, industry would adjust by reducing energy use and substituting other factors of production for energy. The economists were especially skeptical that the government, even in the guise of a quasi-independent SFC, would manage the program efficiently. They further argued that even if it was true that the market price did not adequately account for the external costs of dependence on oil imports, and thus that alternatives to these imports deserved some kind of "national security premium," tax credits would be a better way than the SFC to apply such a premium.

The SFC extended support to several projects in its first year of operation. These included a syngas project and a couple of shale- and coal-to-liquids projects. The projects proceeded well in the sense that the projected schedule and costs were achieved. The SFC eventually demonstrated the technical feasibility, project cost, and environmental impacts of several key synfuels technologies. The problem was that the world oil price, instead of continuing to rise as had been expected, actually declined. The situation can be summarized as follows (with all costs expressed in 1990 dollars):

Actual cost of oil in 1980	$40/barrel
Expectation in 1980 of cost of oil in 1990	$80–100/barrel
Cost of producing synthetic oil	$80–90/barrel
Actual cost of oil in 1990	$20/barrel

Whatever the original purpose of the synfuels program, the reality of declining oil prices meant that synfuels were not needed. Thus, to the public and the critics the program was a failure. A more charitable view of the synfuels effort is that it was an "insurance policy." If the world price of oil had continued to rise, learning about the technical feasibility, cost, and environmental impacts of synfuels technologies would have permitted much more rapid deployment. Because the price of oil did not go up, synfuels production was not needed, but an important risk had been hedged nevertheless. The question from this perspective is whether the insurance could have been purchased more cheaply.

Two criticisms of the program have particular merit. The first is that energy conservation and demand reduction efforts were then and are still today a more cost-effective way to reduce dependence on expensive oil imports. The second criticism is more far-reaching. The synfuels strategy was largely based on building capacity to produce synthetic fuels. In the public and congressional debates, there was tremendous support for adopting a production goal – producing two million barrels per day of oil equivalent by 1990 – independent of the cost of production and the oil price that might prevail in the marketplace. The important lesson that should be learned from the synfuels experience is that a demonstration program of any kind (whether it is for synfuels, or windmills, or photovoltaics) that is based on quantitative goals, independent of the cost and price of alternatives, is tremendously vulnerable to unanticipated price movements. A striking manifestation of this error at the time was a paper produced by Exxon that presented the industry rationale for the synthetic fuels program. The paper projected future energy supply and demand and the value of synthetic fuels. But nowhere in the paper was there any mention of the future market price of oil or the relative cost of producing synthetic fuels.

An insurance program designed to demonstrate the technical feasibility, environmental effects, and costs of alternative technologies can be justified on the basis that it will allow industry to deploy new technology faster if market conditions warrant. But it is a mistake for the government to embark upon a program designed to meet quantitative targets, independent of the prices that prevail in the market. If for one reason or another (e.g., national security, environmental benefits) the government wishes to subsidize an uneconomic energy alternative, it is always better to use indirect incentives like tax credits, which can be adjusted, than to embark on a direct production program that requires massive commitment of public capital.

There is one more lesson which may not be appealing but should not be forgotten: successfully selling a program to Congress requires devoting attention (if not large quantities of money) to other interests in order to attract the necessary political support.

13

Fuel Cells for Automobiles

In this chapter we discuss fuel cells, an exciting energy technology that many hope will become the (environmentally benign) successor to the internal combustion engine for automobile propulsion. Our study of fuel cells illustrates once again a recurring theme of this book – the importance of properly specifying the system boundary when making technology comparisons. The fuel cell case also reveals several important issues that arise in R&D project management.

As societies around the world become increasingly aware of the environmental consequences of energy supply, distribution and use, there is an understandable wish to invent and deploy new technologies that avoid the costs, both environmental and economic, of the technologies in use today. The desire to find something "new," that does not have the drawbacks of what is here now and familiar, is invaluable because it is the fundamental driving force of innovation. But good intentions are not the same as successful outcomes, and it is important to insist on disciplined analysis of the technical, economic, and environmental aspects of a new technology before launching expensive new initiatives. This is true for entrepreneurs thinking about starting a new company around a new technology, for an established company considering an expensive new R&D program, or for a government agency considering adopting a new tax, regulatory, or technology development program.

One of the biggest targets in the search for a qualitatively more attractive energy technology is the automobile. Is it possible to find an alternative to the conventional gasoline-fueled internal combustion engine (ICE) for private automobiles that would have a significantly lower environmental burden while maintaining the same level of performance for the consumer at comparable cost?[1] Several different types of propulsion system have been suggested, including diesel combustion, electric vehicles, hybrid internal combustion/electric vehicles, and, most recently, fuel cells – the subject of this chapter. The federal government has launched a major cooperative effort with industry, referred to as the "Partnership for Next Generation Vehicles" (PNGV), to explore alternative vehicle concepts, and is spending approximately $300 million per year through a consortium with the automobile manufacturers called US CAR.[2]

---

[1] Mass transportation presents another set of alternatives that should be considered in this context.

[2] On January 9, 2002, Spencer Abraham, the Bush Administration's new Secretary of Energy, announced that the PNGV program would be replaced by the FreedomCAR cooperative automotive research program with the same U.S. CAR consortium of auto companies that participated in the PNGV program. The FreedomCAR program changed the focus from near-term demonstration of fuel cell vehicles to the goal of developing infrastructure and technologies for hydrogen-based fuel cell vehicles. Unquestionably part

To put forward a plausible alternative to the existing transportation system is a major challenge. One must imagine an entirely new infrastructure to produce and distribute fuel or energy to a dispersed vehicle fleet. One must consider the design and cost of the new infrastructure, as well as the important question of how society might make the transition from the existing situation to the new. These are not easy problems to address analytically, and they are much harder still to solve in reality.

Today the new technology that is capturing most attention is the fuel cell. A fuel cell is a nonrechargeable battery: fuel and oxidant react in an electrochemical cell and produce direct current (dc) electricity. Several major auto companies, notably including Ford, General Motors, and Daimler Chrysler, have launched major fuel cell programs, evidently because of the potential they see for this technology to compare favorably, in large scale application, with conventional internal combustion engines from a performance, cost, and environmental point of view. The enthusiasm for fuel cells is based on the judgment that this technology can power automobiles more efficiently and with less environmental insult. In order to assess this possibility, it is necessary to examine not just the fuel cell power module itself, but all the other sub-systems that are required to make the overall system work to provide the desired transportation service. Our task in this chapter is to analyze a fuel cell-based transportation system in relation to the alternatives. Appreciating how complex the analysis is prepares us to address other proposals for major infrastructure change.

TECHNICAL ASPECTS OF FUEL CELLS

A fuel cell acts like an unrechargeable battery that is capable of producing direct current electricity as long as fuel and oxidant are supplied to the anode and cathode respectively. Practical fuel cells today operate with hydrogen fuel (see Figure 13.1), although it is possible to build fuel cells that operate on other types of fuels.

If a single fuel cell module does not produce the desired voltage, a group of them can be ganged together in "stacks" to realize the desired electrical characteristics (see Figure 13.2).

The most promising type of fuel cell for automotive operation uses a polymer exchange membrane (PEM) as an electrolyte. The advantage of the PEM fuel cell is its low operating temperature, about 80°C.

In principle the fuel cell has several advantages over an ICE. First, when the fuel cell operates at its design condition of fixed voltage, V, and current, I, it is quite efficient, producing power $P = VI = I^2 R$ (where R is the resistance of the system). Moreover, because the fuel cell produces direct current, and therefore can provide power to the wheel on demand through a direct drive power train, idling loses (which represent about 50% of all ICE losses in a typical drive cycle) are avoided.

of the motivation for this change was the desire to replace an initiative closely associated with the Clinton Administration with a new Bush Administration initiative. But the shift also reflects some of the difficulties discussed in this chapter in realizing fuel cell powered automobiles. In practice, the new FreedomCAR program set a much longer time horizon for introduction of fuel cell vehicles than the PNGV program.

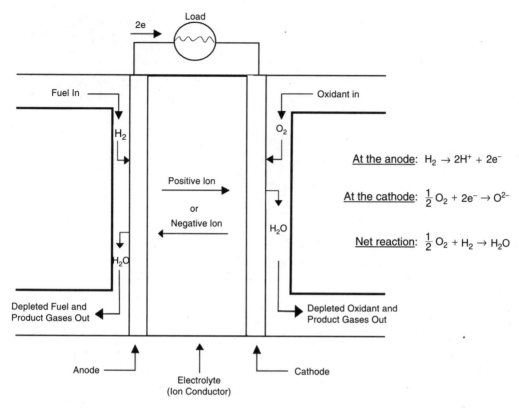

Source: Reprinted from J. H. Hirchenhofer et al, *DOE Fuel Cell Handbook (4th ed.)*, DOE/FETC-99/1076, November 1998.

Figure 13.1. Schematic of a hydrogen-fueled fuel cell.

The theoretical efficiency of a fuel cell is not governed by the Carnot efficiency that determines the performance of a heat engine operating between a high temperature heat source and a low temperature sink for waste heat. The fuel cell offers an interesting alternative to heat engines because different efficiency limits apply.

Second, the fuel cell has lower vehicle emissions. With hydrogen fuel there are no NO_x, CO_2, particulate or volatile organic compound emissions. The only product is water. Of course, one must also be concerned with emissions associated with the production of the hydrogen.

Third, fuel cell vehicles are quiet and, depending on the fuel system, emit less odor. Both features are very important from the viewpoint of consumer satisfaction.

But the fuel cell vehicle also has significant disadvantages. First, the reliability of fuel cells under realistic operating conditions over a period of many years is unproven. Second, the costs of producing fuel cells in volume and of maintaining them in the field are not known. Third, the performance of the fuel cell depends critically on the effectiveness of the catalyst that determines the reaction rates at the electrodes. All effective catalysts are easily poisoned by small amounts of impurities, notably CO and SO_2. Fourth, the hydrogen fuel cell vehicle confronts a difficult design issue with respect

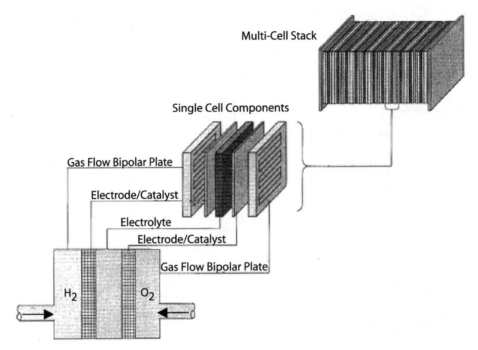

Multi-Cell Stack

Single Cell Components

Gas Flow Bipolar Plate

Electrode/Catalyst

Electrolyte

Electrode/Catalyst

Gas Flow Bipolar Plate

H_2 O_2

Source: Adapted from J. H. Hirchenhofer et al, *DOE Fuel Cell Handbook (4th ed.)*, DOE/FETC-99/1076, November 1998.

Figure 13.2. Expanded view of a basic fuel cell repeated unit in a fuel cell stack.

to fuel carriage and range. Use of hydrogen as the on-board fuel means either that the fuel must be stored in gaseous form, in which case the range will be limited compared to liquid fuel carriage (and problems related to storage, including hydrogen embrittlement of materials, will have to be solved), or, if a liquid fuel such as methanol or gasoline is carried on-board instead, the problem of reforming the liquid fuel into hydrogen on board must be confronted. Finally, because the fuel cell must function at above-ambient temperature, an auxiliary power source is needed to achieve start-up from cold conditions.

THE FUEL CELL "SYSTEM" PROBLEM

At first glance it appears that fuel cell (FC) powered vehicles should be much superior to ICE vehicles on the basis of energy use and emissions. However, any conclusion about the relative merits of the two also requires comparing the ICE gasoline infrastructure to the hydrogen infrastructure on which a fuel cell-powered fleet would run. Where would the hydrogen come from? How much would it cost? Would the fleet work best with hydrogen stored on board or should a liquid fuel (gasoline, methanol, or ethanol) be carried and reformed on board? And how would the use of these liquid fuels in FC systems compare with the use of the same fuels in ICE vehicles instead?

If we wish to do the analytic comparison properly, we must consider, for each engine type, various types of fuel.

ICE engine	FC engine
– Gasoline	– Hydrogen
– Methanol	– Methane
– Ethanol	– Methanol
– Methane	– Ethanol
– Diesel	– Gasoline

A thorough "system" analysis would compare each of these alternative vehicle transportation systems from the source of the hydrocarbon fuel to the point of final energy use, that is, from " the well" to "the wheel." The analysis would compare the technical performance (range, cold start, acceleration, and noise), the economics (initial acquisition cost, operating and maintenance cost, and life-cycle cost), and the environmental effects (emissions of CO_2, NO_x, particulates, and volatile organics). The assessment would include emissions both from the vehicle itself and from the process of bringing the fuel to the automobile's power plant in usable form.

The first comparison is between the ICE (or diesel compression engine) and the FC drive trains. The comparison is shown in the two diagrams below on a common basis: the number of energy units required to move a vehicle of a given weight through a given distance of a specified drive cycle (e.g., the kilojoules required to move a 2,000 kg car through 1 kilometer of an urban drive cycle.)

For an internal combustion engine:

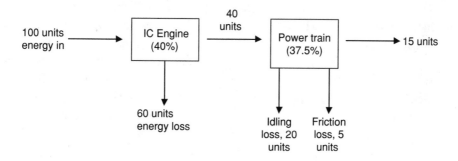

For an equivalent fuel cell/direct drive vehicle (requiring a 65 kWe fuel cell stack):

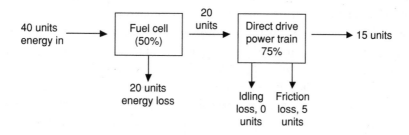

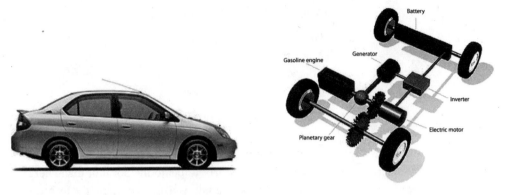

Source: http://www.toyota.co.jp/IRweb/corp_info/eco/advanced_hybrid04.html.

Figure 13.3. The Toyota Prius.

It is immediately apparent that the fuel cell advantage is due mostly to the more efficient direct drive power train and to a lesser extent to the higher efficiency (50% versus 40%) of the engine energy conversion. Thus, as we proceed with our system comparison, we should keep in that an ICE vehicle with a different power train could be much more efficient than the conventional ICE vehicle configuration. The identification of key design features that, if modified, would significantly improve the performance of a system is an important aspect of the systems analysis process.

There is in fact a class of hybrid vehicles that captures the majority of idling losses and hence achieves a sizeable part of this performance improvement. These hybrid vehicles combine an ICE/conventional power train system with a direct drive electric motor with batteries for storage. The vehicle control system continuously chooses between using the ICE to power the car and charge the batteries and using the electric motor for direct drive. This is a formidable engineering system concept that should compete well against both FC and conventional ICE vehicles. Honda and Toyota now produce cars that operate under this principle (the Toyota Prius hybrid is shown in Figure 13.3).

If the comparison of the two process diagrams shown above was all there was to this story, matters would be quite straightforward: the FC would have the advantage over the ICE with regard to energy efficiency and therefore also with respect to emissions on a per mile basis. The only remaining issue would be the relative cost of the two systems for comparable performance, and this in turn would depend on the capital cost of the FC, the relative cost of fuel, and the reliability of FC operation in normal use.

There is, however, a major complication. The ICE system employs gasoline as fuel with a production and distribution system that is relatively energy efficient and has no major emissions problems between the well and the wheel. The FC system, on the other hand, uses hydrogen as fuel, and the hydrogen must be produced and distributed somehow. There are four possibilities for hydrogen production: electrolysis of water using solar energy, electrolysis of water using nuclear energy, reforming ethanol produced from biomass, or reforming natural gas or a petroleum derivative. Producing hydrogen with solar or nuclear energy would decrease greenhouse gas emissions considerably, but at least for now the cost in both cases appears prohibitively high. The biomass avenue may be attractive in the future if it becomes possible to produce large amounts of ethanol at low cost from cellulosic biomass, as discussed in Chapter 2.

The most realistic source of hydrogen at present is petroleum or natural gas. Some of the most likely possibilities are indicated in the flow diagram below:

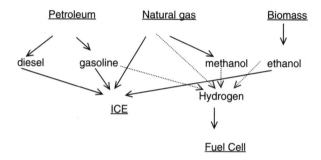

If oil or gas feedstock is employed to produce the hydrogen for the fuel cell, a chemical conversion step is needed. As already noted, the conversion is called reforming – stripping the hydrogen from the hydrocarbon. This step leads unavoidably to a carbon by-product as well. For example reforming methane (the most likely source of hydrogen) at the very least will produce one mole of CO_2 for every 3 moles of hydrogen:

$$CH_4 + \frac{1}{2}O_2 \rightarrow CO + 2H_2$$

$$CO + H_2O \rightarrow CO_2 + H_2$$

$$\overline{\phantom{CH_4 + \frac{1}{2}O_2 + H_2O \rightarrow CO_2 + 3H_2.}}$$

$$CH_4 + \frac{1}{2}O_2 + H_2O \rightarrow CO_2 + 3H_2.$$

We next compare the CO_2 emissions for the ICE and FC systems.

For the gasoline internal combustion engine, the heat of combustion, $\Delta H = -5116$ kJ/mole. Thus, for an energy input of 100 kJ to the ICE the moles of gasoline required $= 100/5115 = 0.0195$ moles. According to the stoichiometry of the combustion reaction

$$C_8H_{18} + \frac{25}{2}O_2 \rightarrow 8CO_2 + 9H_2O$$

each mole of octane burned will produce 8 moles of CO_2. Thus we have 8×0.0195 moles $= 0.1564$ moles of CO_2 produced per 100 kJ of energy input.[3]

For the energetically equivalent hydrogen fuel cell, 40 kJ of hydrogen are needed, as discussed previously. For the production of hydrogen, we again consider the case of methane reforming. The amount of methane energy that is required will depend on the reformer efficiency, ϕ, defined as the ratio of the hydrogen energy produced to the

---

[3] In the Appendix to this chapter, the CO_2 emissions from gasoline ICE engines are compared with those from diesel compression engines and compressed natural gas vehicles.

Table 13.1. Hydrogen fuel cell system methane requirements and CO_2 emissions

Reformer efficiency, ϕ	CH_4 energy (kJoules)	Moles CH_4	$\Delta CO_2(FC)/\Delta CO_2(IC)$	$\Delta E(FC)/\Delta E(IC)$
1	40	0.0499	0.32	0.40
0.9	44	0.0549	0.35	0.44
0.8	50	0.0623	0.40	0.50
0.7	57	0.0711	0.45	0.57
0.6	67	0.0835	0.53	0.67
0.5	80	0.0998	0.64	0.80

methane energy input to the reformer. The latter is given by the heat of combustion

$$CH_4 + 2O_2 \rightarrow CO_2 + 2H_2O \quad \Delta H = -802 \, \text{kJ/mole}.$$

If the efficiency of the reformer is, say, 80%, the number of moles of methane required as input to the reformer is

$$40(\text{kJ}) \times \frac{1}{0.8} \times \frac{1}{802} \left(\frac{\text{moles}}{\text{kJ}} \right) = 0.0623 \, \text{moles},$$

and 0.0623 moles of CO_2 will be produced by the FC system, that is, about 40% of the CO_2 emissions from the gasoline powered ICE system (here we have also assumed no CO_2 emissions from the gasoline production and distribution infrastructure.)

The relative energy requirements and CO_2 emissions of the ICE and FC systems are shown in Table 13.1 and Figure 13.4 as a function of the reformer efficiency.

What conclusion should be drawn? The estimated energy efficiency and environmental impact (measured here by CO_2 emissions) involves the entire system from well to wheel, and not just the performance of one sub-system – the fuel cell power plant.

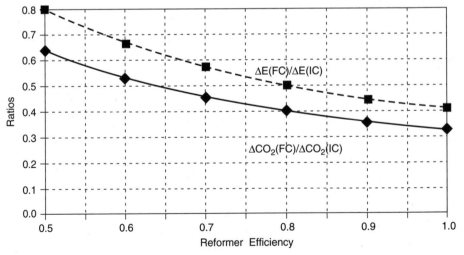

Figure 13.4. Energy efficiency and CO_2 emissions of hydrogen fuel cell system relative to gasoline internal combustion engine.

Moreover, while we have focused in this example on reformer efficiency, the quantitative estimate of relative overall performance will depend upon many other technical parameters besides. The choice of appropriate values for each such parameter can be based on (1) current practice, or (2) projected best practice based on technical or economic criteria.

For example, to avoid having to create an entirely new infrastructure for the distribution of hydrogen fuel, we can consider the use of methanol as a liquid intermediate product between the natural gas at the "well" and the hydrogen gas feed to the fuel cell. The energy efficiency and associated CO_2 emissions of the off-vehicle part of the fuel cycle will be significantly influenced by the design of the methanol plant, which (presumably) will produce methanol by direct oxidation:

$$CH_4 + \frac{1}{2}O_2 \rightarrow CH_3OH.$$

On an ideal basis this reaction releases no CO_2 and is exothermic, so if the excess heat is used productively there should be no external environmental charge for the conversion step. The monetary cost of the conversion would of course be reflected in the product cost of the methanol. As a practical matter, however, some CO_2 is formed in the process, with the exact amount depending upon the technology vintage used in the methanol plant and the process conditions, which will be dictated by plant economics. Other design challenges confronting a methanol fuel supply infrastructure include the very high toxicity and corrosiveness of methanol.

FUEL CELLS AND THE CHALLENGE OF MANAGING GOVERNMENT R&D PROGRAMS

The economic competitiveness of fuel cell vehicle systems hinges on three key cost factors:

1. The capital cost of the fuel cell stack – importantly influenced by the equipment lifetime.
2. The operating and maintenance costs – affected by the reliability of operation of the fuel cell system in the field.
3. The cost of obtaining the fuel cell feed – today assumed to be hydrogen – in a practical manner.

As mentioned above, the U.S. government is working with several automobile companies to develop a practical and economical fuel cell passenger car transportation system. It is not yet clear whether fuel cell technology will prove superior to other alternatives, notably IC/electric hybrids, advanced diesel engines, or compressed natural gas vehicles. The eventual outcome will depend on three factors: technical advances as yet unknown; the demonstrated cost of ownership of the systems in field use; and the direction of environmental regulation.

It is perfectly reasonable for the Department of Energy to mount a research, development, and demonstration program with industry with the objective of advancing fuel cell technology and demonstrating that fuel cell systems for propulsion can be cost competitive. Advocates of government-supported energy technology programs typically call for two kinds of programs: technology programs to improve performance and programs to demonstrate system performance and cost in a practical setting. Frequently, concern over the slow pace of commercialization leads advocates of a particular technology to urge ever more aggressive and costly government support programs. We have already encountered this tendency in the chapters on wind and synthetic fuels. A particularly expensive approach is to advocate a government supported "buy-down" program that seeks to drive down the unit capital cost by exploiting the benefits of technological learning and manufacturing economies of scale. Careful arguments in support of government supported buy-down efforts can be found in two recent reports published by the President's Council of Advisors on Science and Technology (PCAST).[4]

We take this as an opportunity to make two important points about managing federal research, development and demonstration (RD&D) programs: (1) it is usually advantageous to divide RD&D programs into phases in order to control costs; and (2) "buy-down" programs always involve a difficult balance between advancing technology and reducing the cost of current-generation technology.

The first point can be illustrated with a highly simplified example. Let us assume that the energy technology RD&D program can be divided into two stages. The first is a R&D phase with a cost C_1 and a probability of success of p_1. The second is a demonstration phase that has a cost C_2 and a probability of success p_2. If we commit to the entire program at the outset, in a unitary approach, then the expected value of the economic loss if the RD&D program fails:

$$L_{\text{unitary}} = (C_1 + C_2)(1 - p_1 p_2),$$

where we have assumed that the probability of success in each phase is independent of the other.

If we take a phased approach, we proceed to phase 2 only if phase 1 is successful. Thus, the expected loss is:

$$L_{\text{phased}} = C_1(1 - p_1) + (C_1 + C_2)p_1(1 - p_2).$$

The difference between the phased and unitary approaches is:

$$L_{\text{phased}} - L_{\text{unitary}} = -C_1 p_1 (1 - p_2) - C_2(1 - p_1).$$

---

[4] President's Council of Advisors on Science and Technology, *Federal Energy Research and Development for the Challenges of the Twenty-First Century*. Report of the Energy Research and Development Panel of the President's Committee of Advisors on Science and Technology (PCAST), November 1997, (see especially Chapter 7), available at: http://ostp.gov/PCAST/pcastdocs93_2000.html; President's Council of Science and Technology Advisors, *The Federal Role in International Energy Innovation*, Report of the Energy Research and Development Panel, August 1999, available at: http://ostp.gov/PCAST/pcastdocs93_2000.html.

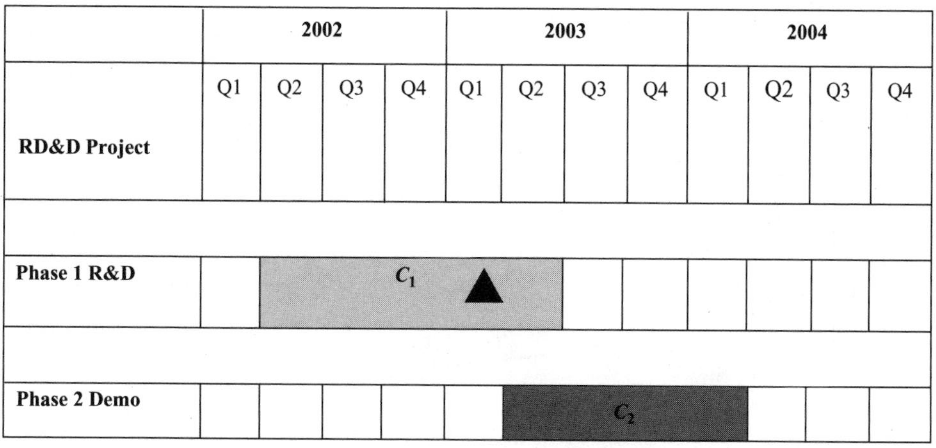

	2002				2003				2004			
	Q1	Q2	Q3	Q4	Q1	Q2	Q3	Q4	Q1	Q2	Q3	Q4
RD&D Project												
Phase 1 R&D		C_1 ▲										
Phase 2 Demo						C_2						

Figure 13.5. Milestone chart for a two-stage RD&D project.

Not surprisingly, the phased approach has a lower expected loss, reminding us that it is wise to break up RD&D programs into phases so that one can proceed to incur additional costs only as technical plans are successfully completed. This is especially important since the later stages of an RD&D program are often much more expensive than the earlier stages. For this reason, too, "crash" programs that set fixed schedules for technical progress should whenever possible be avoided in favor of proceeding as rapidly as successful events allow.

Program managers, whether in the public or private sector, frequently use simple program milestone charts that allow progress to be assessed against planned performance, schedule, and cost in a straightforward manner. Figure 13.5 shows such a chart for the case of the two-stage RD&D project imagined above. The triangle in the chart is the milestone at which project review takes place. The placement of the triangle in this case implies a milestone evaluation towards the end of the first phase in order to determine project progress and to win approval to proceed to the second phase.

This simple example should not be taken too literally. R&D planning is more complex than the example suggests. The probability of success in each phase is influenced by the amount of resources that are allocated to that phase, and perhaps by the expenditures in previous phases as well. The example also suggests that adding additional phases will always lower the expected loss. But this is not so; some risk of failure will remain and adding many phases may create confusion and adversely affect progress.

The second point concerns the choice facing the government RD&D program manager between, on the one hand, spending available resources on a demonstration program to buy fuel cell units with specified performance characteristics based on present technology and, on the other hand, investing money and time in an effort to improve the technology. The fuel cell example clearly illustrates this dilemma. Let us assume that with today's technology the capital cost of a fuel cell is K \$/kWe and the lifetime of the fuel cell is N years. As shown in Chapter 3, if the cost of capital is r per year, the annual

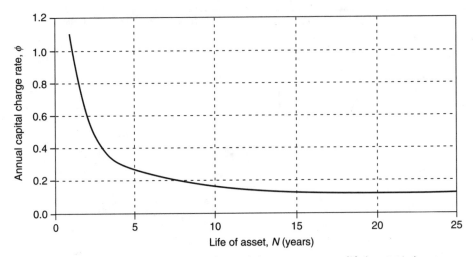

Figure 13.6. Dependence of annual capital charge rate on asset life ($r = 0.1/\text{yr}$).

capital charge rate, ϕ, is given by

$$\phi = r \frac{(1+r)^N}{(1+r)^N - 1},$$

and the capital component of the fuel cell electricity cost (in cents per kilowatt hour) is

$$e_c(\text{cents/kwhr}) = \frac{K \times 100}{8760\,L}\phi = \frac{100\,K}{8760\,L}\frac{r(1+r)^N}{(1+r)^N - 1},$$

where L is the fuel cell system capacity factor.

Next, suppose that early models of the fuel cell system have a lifetime of two years (or about 17,000 hours of continuous operation) and that a technology development program costing C_1 would extend this lifetime to five years. If $r = 0.1/\text{yr}$, this would reduce ϕ from 0.576 to 0.264 – a factor of about 2.18 (see Figure 13.6).

An alternative to spending money on a technology development program to extend fuel cell lifetime would be to use government procurement to drive the production cost down the learning curve. As discussed in Chapter 3, a useful rule of thumb is that for each doubling of the cumulative output of a product the production cost declines by a 'learning' factor, f. Thus, the unit capital cost of the fuel cells in the zth production lot, $K(z)$, is

$$K(z) = f^n K_1,$$

where

$$z = 2^n,$$

and K_1 is the unit capital cost of the fuel cells in the first production lot.

How much might it cost to reduce the unit cost by a factor of 2.18? If $f = 0.9$ then the value of n required is between 7 and 8 (the formula gives $n = 7.4$). At $n = 8$ the cumulative ouput of production lots is $z = 2^8 = 256$, which, depending upon lot size, is quite a large number of units. (Of course, the number would be significantly smaller for more rapid learning, i.e., smaller f).

It is instructive to calculate the cost to the government of buying its way down this learning curve. The total cost of producing $z = 2^n$ lots, TC(z), is

$$\text{TC}(z) = \sum_{n'=0}^{n-1} K_1 f^{n'} 2^{n'} q = K_1 q \frac{(2f)^n - 1}{2f - 1},$$

where q is the lot size.[5]

As the buy-down cost curve in Figure 13.7 shows, producing 256 lots would drive down the unit capital cost by a factor of about 2.3 and would require a cumulative expenditure equal to about 136 times the cost of the first lot. For illustrative purposes, assume that the cost per kilowatt of the fuel cells in the initial lot, K_1, is \$3,000/kwe, the capacity of a fuel cell stack is 25 kWe, and the lot size is 20. In this case the buy-down program would cost about \$200 million – possibly much more than the cost of a technology development program to extend the fuel cell lifetime. The fuel cell enthusiast might argue that both should be done. But if so, the two programs would need to be undertaken in series rather than in parallel since the fundamental process underlying the learning curve is that of becoming more efficient by doing the same job repeatedly.

This example illustrates the dilemma project managers in both government and industry continually face between spending R&D dollars to improve system performance (but perhaps also increasing the production cost of the device) and freezing the design and the performance it embodies in order to realize production efficiencies and hence lower cost.

The example given here also raises the policy question of when it is appropriate for government to invest in research and development on commercial products and processes. This important topic is explored in greater depth in Chapter 15.

---

[5] In the continuous limit,

$$K(z) = K_1 \left(\frac{q}{x}\right)^{\frac{1-f}{\ln 2}}.$$

If we seek to reduce the capital cost by a factor b, the equation $b = K(x_0)/K_1(q)$ determines the size of the production run x_0 required. The total cost of this production run is given by:

$$\text{TC}(x_0) = \int_q^{x_0} K(x)dx = K_1 \int_q^{x_0} \left(\frac{q}{x}\right)^{\frac{1-f}{\ln 2}} dx = \frac{K_1}{1 - \left(\frac{1-f}{\ln 2}\right)} \left[x_0^{1 - \frac{1-f}{\ln 2}} - q^{1 - \frac{1-f}{\ln 2}}\right].$$

In the limit $f \to 1$ both the discrete and continuous expression for the total cost are the same:

$$\text{TC}(x_0) \to K_1 q \left[\frac{x_0}{q} - 1\right] = K_1 q \left[2^{n_0} - 1\right].$$

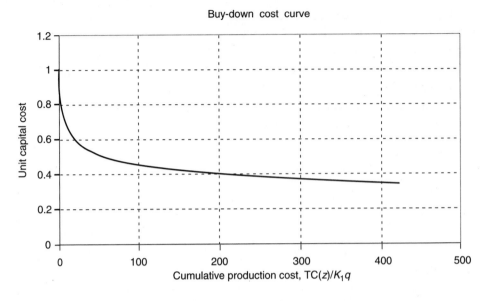

Figure 13.7. Relationship between unit capital cost and cumulative production cost ($f = 0.9$).

n	Cumulative output (number of lots) $z = 2^n$	$K(z)/K_1$ $= f^n$	Total cost, $TC(z)/K_1 q$
0	1	1.00	0.00
1	2	0.90	1.00
2	4	0.81	2.80
3	8	0.73	6.04
4	16	0.66	11.87
5	32	0.59	22.37
6	64	0.53	41.27
7	128	0.48	75.28
8	256	0.43	136.50
9	512	0.39	246.70
10	1024	0.35	445.06

APPENDIX TO CHAPTER 13

Greenhouse Gas Emissions from Hydrocarbon Fuels

There is a great deal of confusion about which hydrocarbon fuel produces the most energy per unit of CO_2 emitted. The fuels most often in question are methane in compressed natural gas vehicles, gasoline in IC engines, and diesel fuel in diesel compression engines. Two factors must be considered: the ratio of the energy content of the fuel to the CO_2 emitted and the efficiency of the fuel in the engine.

Table 13.A.1. Calculation of heats of combustion for straight-chain alkanes (kJ/mole)

(1)	(2)	(3)	(4)	(5)	(6)	(7)
n	$\Delta H_f^o(H_2O)$	$\Delta H_f^o(H_2O) +$ $\Delta H_f^o(CO_2)$	$(1) \times (3)$	$(2) + (4)$	$\Delta H_f^o(C_nH_{2n+2})$	$(5) - (6)$
0	−241.82	−635.33	0	−241.82	0	−241.82
1	−241.82	−635.33	−635.33	−877.15	−74.81	−802.34
2	−241.82	−635.33	−1,270.66	−1,512.48	−84.68	−1,427.8
3	−241.82	−635.33	−1,905.99	−2,147.81	−103.89	−2,043.92
4	−241.82	−635.33	−2,541.32	−2,783.14	−126.15	−2,656.99
5	−241.82	−635.33	−3,176.65	−3,418.47		
6	−241.82	−635.33	−3,811.98	−4,053.8	−167.19	−3,886.61
7	−241.82	−635.33	−4,447.31	−4,689.13		
8	−241.82	−635.33	−5,082.64	−5,324.46	−208.45	−5,116.01

We present a simple analysis of this question using n-octane as a reference fuel for gasoline.[6] First, we calculate the heats of combustion for straight chain alkanes according to the chemical reaction:

$$C_nH_{2n+2}(g) + \frac{2n+1}{2}O_2(g) \rightarrow nCO_2(g) + (n+1)H_2O(g).$$

We use the formula below and tabulated standard heats of formation. The results are given in Table 13.A.1 and Figure 13.A.1.

$$\Delta H_{combustion}(C_nH_{2n+2}) = n\Delta H_f^o(CO_2) + (n+1)\Delta H_f^o(H_2O) - \Delta H_f^o(C_nH_{2n+2}).$$

The tabulation shows close to a straight line relationship between the heat of combustion and the number of carbon atoms in the straight chain alkanes. A good approximation is:

$$\Delta H_{combustion}(C_nH_{2n+2}) = -186.1 - 616.2n \text{ kJ/mole}.$$

With this approximation the heat of combustion of any alkane can be estimated. A comparison of the three fuels, methane, gasoline, and diesel, is shown in the following table:

n		$\Delta H_{comb}(C_nH_{2n+2})$ (kJ/mole)	CO_2(mole/ mole fuel)	$\Delta H_{comb}(C_nH_{2n+2})$ (kJ/g fuel)	$\Delta H_{comb}(C_nH_{2n+2})$ (kJ/g CO_2)
1	methane	−802.3	1	−50.15	−16.72
8	gasoline	−5,116.0	8	−47.59	−13.22
16	diesel	−10,045.9	16	−44.45	−13.08

[6] Gasoline produced in a "straight run" refinery is likely to have a C/H ratio similar to n-octane. If the gasoline is "reformulated," there will be considerable amounts of olefins and aromatics that will reduce the C/H ratio by 5% to 10%; this leads to a considerable modification to the results presented here.

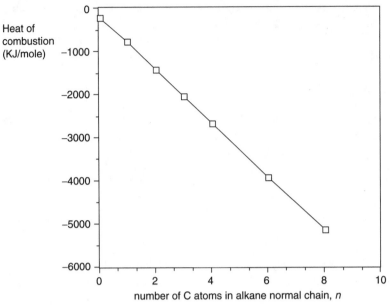

Figure 13.A.1. Heat of combustion of straight chain alkanes.

This table shows that our model gasoline and diesel compounds have about the same energy content per gram of fuel and comparable energy content per gram of CO_2 emitted, and that methane has a considerable advantage in energy per gram of CO_2 emitted.

However, fuel is not sold on a weight basis. Liquid fuels are sold on a volume basis, so the density of the liquid fuel matters. Diesel is about 15% denser than gasoline so on a "per gallon basis," diesel has an energy advantage.

Diesel engines are also more efficient than ICE engines by about 15%, so we expect diesel engines to be about 15% more efficient than gasoline engines on a per weight basis and about 30% more energy efficient on a volume basis. This is quite an advantage. As fuel prices (always expressed on a volume basis) increase, the advantage of diesel-powered vehicles over gasoline-powered vehicles also increases. In Europe, where fuel prices are three or four times those in the United States, there is strong demand for diesel vehicles.

As we have seen, higher engine efficiency also gives diesels an environmental advantage in terms of CO_2 emissions. Diesels also, however, have an important environmental disadvantage relative to gasoline ICEs in their production of particulates. Diesel particulates, odor, and noise, combined with relatively low fuel prices, have slowed the introduction of diesel passenger vehicles in the United States.

14

Energy Models and Statistics

Sensible energy planning by governments and corporations requires assumptions about future energy prices and the amounts of energy that will be produced and consumed. This planning also requires accurate and comprehensive data on current and previous energy activities, so that decisions will be as informed as possible. Good historical data and comprehensive energy models that forecast the future are critical for both private industry and government. In this chapter, we first discuss the kinds of statistics that are available, most of them collected by the federal government, and then the efforts to project the characteristics of future energy activity, both in the United States and in the world as a whole.

Of course, the historical data contain a good deal of random error and are not necessarily accurate. Analysts must use statistical tools to determine what can be inferred from the data. We discuss some of these tools in the second part of the chapter. Statistical analysis is an important specialized field and here we do little more than to illustrate the kind of analysis that can be undertaken and the types of questions that it raises.

An energy forecast seeks to generate a projection of future quantities and prices of various types of energy – both on the supply side and the demand side – based on mathematical models of energy markets and historical data. Such projections require assumptions about many different factors, for example:

- the level of economic activity;
- the nature and extent of future regulations that will influence the cost of producing or utilizing different types of energy;
- the conditions in international markets;
- the cost of capital and of technology;
- technological changes that will influence the cost of producing energy or the efficiency of energy use and substitution.

There are two independent economic processes that together determine energy prices and quantities. The first is the demand for energy by consumers. The *demand curve* describes the quantity, q, of a particular type of energy desired at a given price, p. Figure 14.1 shows a schematic demand curve, $q_d(p)$, with $q_d'(p) < 0$, since consumers use less of a commodity as the price increases (or, equivalently, more of the commodity as the price decreases).

The second process is that of supply. The supply curve describes the quantity of energy, $q_s(p)$, that is offered at a given price, with $q_s'(p) > 0$ because producers offer more energy as the price increases.

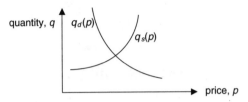

Figure 14.1. Supply and demand curves.

In general it is difficult to obtain an exact picture of either the supply or the demand curve. The reason is that empirical observation reveals only the market clearing price and quantity, that is, the intersection of the supply and demand curves.

The responses of supply and demand to price are characterized by the respective *price elasticities* of supply and demand, defined as

$$\text{price elasticity}, \ e = \frac{d(\ln q)}{d(\ln p)} = \frac{p}{q} \frac{dq(p)}{dp}$$

Frequently, the supply and demand curves are approximated by the following functional forms:

$$q_d(p) = \text{const.} \, p^{-\alpha}$$
$$q_s(p) = \text{const.} \, p^{\beta},$$

where $\alpha, \beta > 0$, and are the price elasticities of demand and supply, respectively.

ENERGY DATA – SOURCES AND UNITS

Macro models of stocks and flows of energy deal in *primary* energy. This is reasonable for coal, oil, and gas, but it is problematical for other types of energy such as hydroelectric and other renewables. In the case of fossil fuels, the primary energy is produced directly and there is a market for the commodity – coal, oil, and gas. In the case of renewables, the primary source of the energy – sunlight, wind, or water – is not a commodity that is produced and traded in markets.

In the United States, energy quantities are typically given in British Thermal Units, or BTUs. A BTU is the amount of energy required to heat one pound of water by one degree Fahrenheit (°F). For large quantities of energy a commonly used unit is the 'quad', equal to one quadrillion BTUs (i.e., 10^{15} BTU).[1]

The official source of energy statistics in the United States is the Energy Information Administration (EIA) of the Department of Energy (DOE). EIA collects and publishes historical data on energy sources and uses. The historical data are important, as we shall see, in understanding trends in energy markets. The EIA uses the data and models to make projections about future energy sources and uses. The EIA was created in 1978 in

---

[1] Other useful units and conversions include: 1 BTU = 1055 joules; 1 barrel of oil (bbl) = 5.8 million BTU; 1 bbl = 0.136 metric tons; 1000 cubic ft (MCF) of natural gas ≈ 1 million BTU; 1 MCF = 0.028 m³.

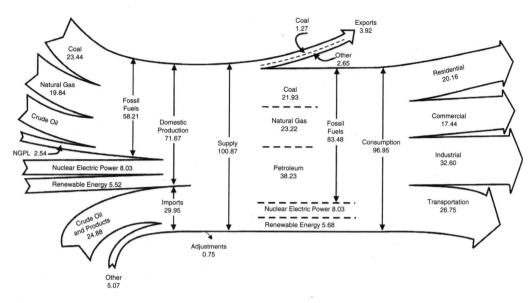

Source: Reprinted from Energy Information Administration, *Annual Energy Review – 2001*, http://www.eia.doe.gov/emeu/aer/
pdf/pages/sec1.pdf.

Figure 14.2. Energy flows in the United States, 2001 (quadrillion BTU).

recognition of the importance of accurate and detailed energy statistics for informed
policy-making and sensible capital investment decisions. Evaluating the desirability of
an investment obviously requires some estimate of future prices. Statistics collected by
the government are a valuable starting point for such evaluations, although government
forecasts are far from perfect, as we shall see, especially the farther into the future they
extend.

PRIMARY ENERGY FLOWS IN THE U.S.

The EIA "energy flow diagram" in Figure 14.2 gives an excellent visualization of current
energy movements in the United States from domestic production and imports to final
use. (The flows are reported in quads.) The chart highlights many important aspects of
U.S. energy consumption:

- First, energy consumption in the United States is large, both in absolute terms and in
  terms of energy use per unit of economic output (GDP). (The 97 quads of primary
  energy consumed in the United States can be compared with total world primary
  energy consumption of about 400 quads. Similarly, U.S. per capita primary energy
  use in 2000 was about 345 million BTU per year, compared with the global average
  per capita consumption of about 65 million BTU per year.)
- Second, Americans are great consumers of oil, both for transportation and in industry. While the United States produces a great deal of oil domestically, it imports about
  60% of its total consumption of oil and petroleum products. This dependence on

Table 14.1. U.S. energy projections, 2000–2020

	U.S. consumption of primary energy (quads)			
	1997 Actual	2000	2010	2020
Petroleum	36.4	38.2	44	49.1
Gas	22.6	22.1	27.7	32.3
Coal	21.3	22.5	25.1	26.6
Nuclear	6.7	7.6	6.7	4.6
Renewables	7	6.7	7.4	8
Total	94.4	98.2	111.3	121

Source: Energy Information Administration, Annual Energy Outlook 2000; reference case.

imported oil means that America's payments to producing countries will increase if the world oil price rises. The United States therefore has a great interest, as do other oil importing nations, in the stability and security of oil-producing countries and regions of the world.

- Third, the United States uses a lot of coal, especially for generating electricity. This coal use is a major source of CO_2, NO_x, and SO_2 emissions.
- Fourth, nuclear and renewable energy sources are each a small part of our energy budget (8% and 6% respectively). Many energy experts (although not necessarily the same ones in each case) believe that the use of these energy sources should be increased because of environmental concerns (from the burning of fossil fuels) and the inevitable depletion of fossil, and especially petroleum, resources. Renewables largely consist of hydroelectricity, wood biomass, and geothermal; there are lesser amounts of solar heating, waste energy use, and wind. In the year 2000 the renewable contribution to primary energy was as follows:

	Amount (quads)
Hydroelectric	3.1
Biomass	2.9
Geothermal	0.3
Solar energy	0.06
Wind energy	0.06
Total renewables	6.4

PROJECTIONS OF U.S. ENERGY USE

Each year the EIA projects energy use in the United States over the next few decades. These projections give an impression of the trends that are occurring in world energy markets as well as assumptions about future economic growth and industrial performance in the United States and overseas. The EIA publication that contains these projections is called the "Annual Energy Outlook" (AEO), which can be accessed on the internet at: http://eia.doe.gov/oiaf/aeo/earlyrelease.

Table 14.2. Nuclear generating
capacity (gigawatts)

Year	U.S.	World
1997	99	351.9
2000	97.5	349.9
2010	94.3	339.6
2020	88.8	278.8

Source: Energy Information Administration,
International Energy Outlook 2002, Table 17.

Table 14.1 gives the projection from AEO 2000 out to the year 2020. The EIA is careful to point out that its projections are simply consistent scenarios that follow from specific assumptions about future economic activity, energy demand, and prices. Nevertheless, these projections are influential in both the policy and business communities. Looking back at previous projections reveals much about how informed opinion has changed and about how changeable this expert opinion is.

The EIA projections contain some significant trends:

- There is continued energy growth over the period. Oil and natural gas meet most of the primary energy growth.
- There is no appreciable increase in the use of renewable energy.
- Nuclear energy undergoes a precipitous decline during this period. Table 14.2 presents the anticipated decline in nuclear capacity in the United States and in the world as a whole.[2]

The EIA also keeps track of how energy is used and projects future use patterns. Table 14.3 presents EIA's projection of U.S. energy consumption by end-use sector. Changes in end-use patterns are gradual and occur only after usage patterns and investment have adjusted to real relative price changes or changes in government regulation.

Finally, the EIA also projects the mix of primary energy that will serve each of the end-use sectors. Table 14.4 gives, as an example, the energy delivered to the industrial sector.

For the United States and other nations, electricity generation is key. Electricity is important for industry and commerce, and in most modern economies the electric power industry requires large investments for generation, transmission, and distribution. Table 14.5 gives the projected primary energy sources for electricity production in the United States.

In the past, the EIA projected that future electricity growth would be fueled mainly by coal. This was based on estimates that coal would cost much less than natural gas or petroleum, which in turn reflected a conservative view of the availability of natural

---

[2] As this book went to press, the EIA issued a revised nuclear energy projection for the United States, reflecting recent trends in nuclear plant operating license extensions and plant capacity upratings, as well as improved operating performance. Based on these trends, the EIA now expects U.S. nuclear capacity to remain roughly constant through 2025. As before, though, no new plant orders are projected during this period.

Table 14.3. U.S. energy consumption by sector

	U.S. primary energy consumption by sector (quads)		
	2000	2010	2020
Residential	19.9	21.7	23
Commercial	16.1	17.8	18.2
Industrial	35.2	39.1	42.2
Transportation	27	32.7	37.5
(Electricity-related losses)	(25.4)	(27.7)	(28.5)
Total primary energy	98.2	111.3	121

Source: U.S. Energy Information Administration, *Annual Energy Outlook – 2000.*

Table 14.4. Projected energy consumption in the U.S. industrial sector

	Delivered energy (quads)		
	2000	2010	2020
Petroleum	9.6	10.4	11.5
Natural gas	9.6	11	12
Coal	2.3	2.4	2.4
Renewable energy	2.1	2.4	2.6
Electricity	3.6	4.5	4.7
TOTAL	27.2	30.7	33.2

Source: U.S. Energy Information Administration, *Annual Energy Outlook – 2000.*

Table 14.5. Projected primary energy use in the U.S. electric power sector

	Primary energy use (quads)		
	2000	2010	2020
Petroleum	0.9	0.5	0.4
Natural gas	4.2	6.6	9.7
Coal	20	22.5	24
Nuclear	7.4	6.7	4.6
Renewable energy*	4	4.4	4.8
TOTAL	36.5	40.7	43.5

* Includes conventional hydroelectric, geothermal, wood and wood waste, municipal solid waste, other biomass, petroleum coke, wind, solar thermal, and photovoltaic.

Source: U.S. Energy Information Administration, *Annual Energy Outlook – 2000.*

Table 14.6. Projections of oil and gas prices

Year	Crude oil[1] 1998 $/bbl	Natural gas[2] 1998 $/MCF
2000	21.19	2.17
2010	21.00	2.60
2020	22.04	2.81

[1] World oil price
[2] Domestic wellhead price

Source: U.S. Energy Information Administration, *Annual Energy Outlook – 2000.*

gas reserves. Similar projections of major increases in coal use were also made for the developing economies, notably China and India, where rapid future economic growth is anticipated. This projection has negative implications for CO_2 emissions. The latest EIA projections, as reflected in Table 14.5, are more optimistic about the price and, by inference, the availability of natural gas to meet electricity growth. Time will tell if these projections are correct.

Key Assumptions

We have emphasized that the EIA energy projections are both necessary and useful. However, the projections depend upon a number of key assumptions that must be borne in mind. Most of the projections presented in this chapter are based on assumptions adopted by EIA in its "reference case" for the AEO 2000 report. The important assumptions are:

- Continued growth of the U.S. economy through the period at an average annual rate of 2.2%. High economic growth is, of course, "good" for energy providers and "bad" for those who want to encourage energy conservation.
- Continued reduction in energy intensity during the period. From 1998 to 2020 real GDP is assumed to increase by 61%, while energy use increases by 27%.
- No visible effect of policy measures to reduce greenhouse gas emissions that might be adopted as a consequence of the Kyoto Agreement.
- Stable oil and gas prices over the entire period (see Table 14.6).
- Continued availability of oil and gas at the assumed prices during the period. In particular OPEC production is assumed to grow from 32 million b/d in 2000 to 56 million b/d in 2020 in the reference case.

These trends are quite different from those presented in previous EIA forecasts. How reliable are the projections? The short answer is, not particularly. Neither the U.S. government not independent energy experts have a very good track record. Consider the past predictions of world oil prices shown in Figure 14.3 and Table 14.7. In 1980, experts estimated that the price of a barrel of oil in 1990, ten years into the future, would

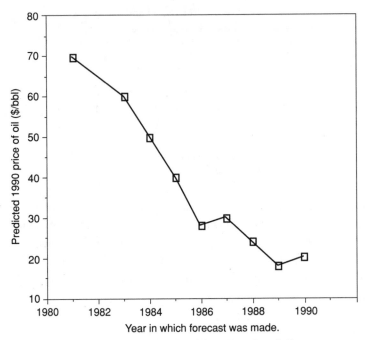

Figure 14.3. Projections of the 1990 price of oil.

be $70; the actual price in 1990 was less than $20. Projections of energy consumption have been scarcely more accurate, as Table 14.7 shows.

The actual world price of oil has not followed a consistent trend, as Figure 14.4 shows. The EIA has consistently underestimated the moderating effect of price on the demand for energy, especially petroleum, and the stimulating effect of price on supply, both through better oil recovery technology and by opening new exploration areas. Indeed, in hindsight the errors of past forecasts are clear:

- In the 1970s, forecasters exaggerated the power that OPEC would have over oil prices, and failed to take account of the sharp reduction in energy use that occurred as consumers and industry responded to higher oil prices by reducing their demand. The demand reduction occurred not only because people used (and wasted) less energy but also because of investments in new, more energy-efficient equipment, for example, more efficient electricity plants and smaller, more fuel-efficient automobiles.

Table 14.7. Previous projections of U.S. energy demand and world oil prices

	Projections for the year 2000	
	Primary energy (quads)	World oil price ($/bbl)
National Energy Plan – 1979	120	100
National Energy Plan – 1981	100	62
Gas Research Institute forecast – 1992	92	35

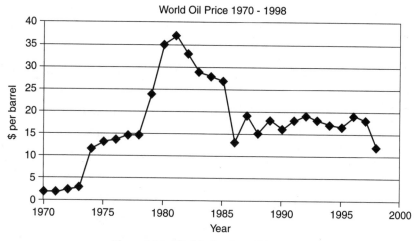

Figure 14.4. World oil prices, 1970–1998.

- In the 1980s, forecasters underestimated the availability of natural gas and overestimated the future use of coal and nuclear energy.
- In the 1990s, the projections were based on an assumption of increased availability of natural gas and oil supplies at moderate prices. As to the reliability of these projections, time will tell.

EIA also projects future world energy use. These projections are contained in the annual EIA "International Energy Outlook." Table 14.8 contains the projection given in the 2002 edition of this report.

World Oil Outlook

Oil is an essential raw material for all economies. Modern societies and emerging economies both depend on petroleum as a transportation fuel, as an essential feedstock for petrochemicals, and as a convenient liquid fuel for power generation. In the short run it is very difficult to substitute other fuels for liquid petroleum.

Oil reserves are not found uniformly throughout the globe. Table 14.9 shows where most of the world's oil reserves and production are found. Certain countries, notably in the Middle East, possess vast reserves that can be produced at relatively low cost. Other countries possess no oil at all and must import to meet their internal needs. Among developed nations, Germany, France, and Japan have essentially no oil. Nor do the developing nations of Central America and the Caribbean, and much of Africa. Several large countries have significant oil reserves and production but still require imports to meet their domestic needs, including Brazil and the United States.

Over time a world market has developed for oil, so that all crude oils produced – such as West Texas Intermediate (WTI) and Nigerian Bonney – are benchmarked to a single world oil price. The price differential of a particular crude relative to the benchmark reflects quality differences and differences in the transportation costs required to bring

Table 14.8. World total energy consumption by region, 1990–2020 (quads)

Regions/Country	History			Projections				Average Annual Percent Change, 1999–2020
	1990	1998	1999	2005	2010	2015	2020	
Industrial Countries								
North America	100.1	112.7	115.7	129.3	140.5	151.3	161.3	1.6
United States	84.2	94.6	97.0	107.6	115.6	123.6	130.9	1.4
Canada	10.9	12.1	12.5	13.7	14.8	15.8	16.7	1.4
Mexico	5.0	6.1	6.1	7.9	10.0	11.9	13.7	3.9
Western Europe	59.8	65.8	66.0	71.5	74.7	77.7	81.5	1.0
United Kingdom	9.3	9.9	9.9	10.7	11.2	11.7	12.2	1.0
France	8.8	10.2	10.3	11.2	11.7	12.3	13.0	1.1
Germany	14.8	14.2	14.0	15.3	15.9	16.4	17.0	0.9
Italy	7.0	8.0	8.0	8.9	9.4	9.9	10.4	1.2
Netherlands	3.4	3.8	3.8	4.1	4.3	4.4	4.6	0.9
Other Western Europe	16.6	19.8	20.0	21.3	22.2	23.0	24.3	0.9
Industrialized Asia	22.8	27.5	27.9	29.7	31.5	33.2	34.9	1.1
Japan	17.9	21.5	21.8	22.9	24.2	25.4	26.6	1.0
Australasia	4.8	6.1	6.2	6.8	7.3	7.8	8.3	1.4
Total Industrialized	182.7	206.1	209.7	230.6	246.6	262.2	277.8	1.3
EE/FSU								
Former Soviet Union	60.7	38.7	39.2	44.1	48.0	53.1	57.1	1.8
Eastern Europe	15.6	11.9	11.2	12.7	13.8	15.2	16.3	1.8
Total EE/FSU	76.3	50.6	50.4	56.8	61.8	68.2	73.4	1.8
Developing Countries								
Developing Asia	51.0	72.9	70.9	92.5	113.9	137.1	162.2	4.0
China	27.0	35.3	31.9	42.9	55.1	68.8	84.4	4.7
India	7.8	11.6	12.2	15.2	18.2	21.8	25.4	3.6
South Korea	3.7	6.9	7.3	9.6	10.7	12.0	13.0	2.7
Other Asia	12.6	19.1	19.5	24.8	29.8	34.6	39.5	3.4
Middle East	13.1	19.1	19.3	22.0	26.3	30.5	34.8	2.8
Turkey	2.0	3.0	3.0	3.4	3.9	4.5	5.1	2.6
Other Middle East	11.1	16.1	16.4	18.7	22.4	26.0	29.7	2.9
Africa	9.3	11.6	11.8	14.0	15.7	18.1	20.3	2.6
Central and South America	13.7	19.4	19.8	22.7	28.3	35.6	43.1	3.8
Brazil	5.7	8.2	8.5	9.4	11.5	14.0	16.8	3.3
Other Central/ South America	8.1	11.2	11.2	13.3	16.8	21.6	26.3	4.1
Total Developing	87.2	123.0	121.8	151.2	184.1	221.3	260.3	3.7
Total World	346.2	379.7	381.9	438.6	492.6	551.7	611.5	2.3

Source: U.S. Energy Information Administration, *International Energy Outlook – 2002*, Table A1.

the oil to market. The world oil market price at any given time depends upon the amount of oil produced and the amount demanded at that time, as well as expectations of future trends in supply and demand.

Major producing nations have attempted to maintain a high world oil price by agreeing to restrict output. A number of oil-producing nations are members of the

Table 14.9. World oil reserves and production

Country	Proved reserves (billion barrels)	1997 production[a] (million barrels/day)
China	24	3.2
India	4.3	0.8
Indonesia	5	1.8
Norway	10.4	3.3
United Kingdom	5	2.8
Russia	48.6	6.1
United Arab Emirates	97.8	2.5
Iran	93	3.7
Iraq	112.5	1.2
Kuwait	96.5	0.9
Saudi Arabia	261.5	9.3
Algeria	9.2	1.4
Libya	29.5	1.5
Nigeria	16.8	2.3
Canada	4.8	2.6
Mexico	40.4	3.4
Venezuela	71	3.5
United States	22.5	9.5
Total World	1020.1	74.3

[a] Includes crude oil, natural gas processing liquids, other liquids, and refinery gain

Organization of Petroleum Exporting Countries (OPEC), which controls a considerable portion of the world's oil reserves and production. OPEC has sought to use this control to wield monopoly power in the marketplace. OPEC member countries include Saudi Arabia, Iran, Iraq, Kuwait, the Gulf States, Egypt, and Libya. Non-Middle Eastern member countries include Nigeria, Venezuela, and Indonesia. Several important oil-producing countries are not members, including Mexico and Canada. States of the former Soviet Union which are important oil producers, notably Russia, Azerbaijan, and Kazakhstan, are also not OPEC members.

On two occasions, in 1973 and in 1978, OPEC unilaterally reduced production, established production quotas, and forced a sharp increase in world market prices. Over time, as often happens in cartels formed to gain market power, the incentive for member countries to cheat and produce above their quota led to a collapse in the price and a return to "normal" market conditions. Importing countries, especially industrial countries in Europe, Japan, and the United States are, of course, interested in encouraging producing countries to maintain high production rates in order to keep world market prices down.

Another concern for importing countries is that much of the world oil supply comes from the Middle East, which is politically highly unstable and subject to considerable violence. In addition to the on-going Arab-Israeli conflict, the region has experienced a war between Iran and Iraq during the 1980s, the Iraqi invasion of Kuwait in 1991, and

Table 14.10. Projected sources of U.S. petroleum supplies

Sources of U.S. petroleum	(Millions of bbl/d)			
	2000	2010	2020	2025
Domestic production[1]	7.78	8.06	8.01	7.96
Net imports[2]	10.42	13.76	17.72	19.79
Total petroleum supplies	18.20	21.82	25.73	27.75

[1] Includes natural gas liquids
[2] Includes imported crude & products
Source: U.S. Energy Information Administration, *Annual Energy Outlook –
2003.*

the Iraqi war of 2003. Several of the major oil producing countries, such as Iran, Iraq, and Libya, have been hostile to the United States and the West, and several other major exporting countries face significant internal political and social challenges, including Saudi Arabia and Kuwait.

The major oil reserves in the Middle East make it a region of immense geopolitical importance. The political instability in the region means that the region and its oil are important national security issues. This national security aspect of oil makes it different from other commodities that we depend on and import, such as coffee. For these other commodities, there are either multiple sources of supply in world markets or substitutes available, or both. In contrast, projecting what may happen to world oil markets – supply, demand, and price – is critical to our national wellbeing.

The U.S. Petroleum Supply Situation

Table 14.10 presents EIA's latest (2003) estimate of future sources of U.S. petroleum supply. Two features are noteworthy. First, oil consumption is projected to continue to increase for the next 20 years. Second, the United States is projected to become increasingly reliant on imported oil. This imported oil will come from OPEC or from other oil-producing countries. Clearly the United States and other oil-importing nations have an interest in stimulating oil production outside of OPEC and the unstable Middle East, and a great deal of effort has been expended in attempting to do this. A notable example is the Caspian Sea region, which is known to have considerable undeveloped resources. The United States has been supporting the construction of pipelines that would bring this oil to market. Russia and Iran understandably have been resisting this expansion because they do not want to lose their historic political and economic control of this region and its resources.

Because there is a world oil market, it does not much matter if one importing country's supply comes from OPEC or from non-OPEC sources. For if, in a given year, country A gets less of its supply from OPEC, the world market adjusts and sends more OPEC oil to country B. Thus, importing countries have a common set of economic and security concerns with respect to world oil supply in both the short-term and the long-term.

ENERGY STATISTICS

We have focused so far on EIA projections. We now turn to a discussion of how the historical data reported by EIA can be used to establish a relationship between two variables in energy markets. Establishing a quantitative relationship between variables is important for constructing the energy models upon which projections are based. But this is not an easy task: the form of the posited relationship may not be correct, other variables may also be involved and vary in an unknown way, and the reported data may include errors of unknown origin and magnitude.

In this section we seek to determine whether two (or more) variables are correlated with each other. The general problem is to ask what functional form will best predict the dependence of a set of dependent variables $\vec{y}$ on a set of independent variables $\vec{x}$. The functional form is generally not specified. This is the central problem addressed by statistical inference and it deserves thorough study. Here we seek only to illustrate the power of this tool for analyzing public policy problems.

We shall consider the simplest case. We examine a set of observations on a single variable y (for example, EIA-reported energy use) that we denote as $\{y_n\}$, and we ask if this set is correlated with another variable x (for example, reported GDP) that we denote as $\{x_n\}$. In particular, we shall assume that a linear relationship exists between the two variables

$$y = \beta x + \alpha + \text{random error.}$$

We ask what is the best estimate we can make of the coefficients (α, β) in the assumed linear relation

$$\hat{y} = \beta x + \alpha.$$

The random error term, hopefully small, is denoted ε and, for a sufficiently large sample, is assumed to average to zero

$$y = \beta x + \alpha + \varepsilon.$$

The average of each quantity is defined by

$$\langle F \rangle = \frac{1}{N} \sum_n F_n.$$

We also use the notation: $\delta F = F - \langle F \rangle$.

Results for One Variable

In the Appendix to this chapter, we show that the best estimate of β is given by:
$\beta = \dfrac{\langle \delta y \delta x \rangle}{\langle \delta x^2 \rangle}$, while α is found from the relation: $\langle y \rangle = \beta \langle x \rangle + \alpha$. A measure of the goodness of fit is given by the "correlation coefficient," r

$$r^2 = \frac{\text{predicted variation}}{\text{total variation}} = \frac{\langle \delta y \delta x \rangle^2}{\langle \delta y^2 \rangle \langle \delta x^2 \rangle}.$$

The analysis can be generalized to more than one independent variable

$$y = \alpha + \beta_1 x_1 + \cdots + \beta_s x_s + \varepsilon.$$

This is discussed in the Appendix.

Confidence Intervals

How likely is a particular observation y_n to lie within a given interval around the value $\hat{y}_n$ predicted by the linear relationship? A surprisingly precise set of statements can be made in answer to this question using statistical inference. Here we take a very simple approach that suggests what is possible.

Note that the deviation between actual and predicted observations is directly related to the random error term ε:

$$d_n = y_n - (\alpha + \beta x_n) = \varepsilon_n.$$

Until now, no assumption has been made about the statistical nature of the random error term. If we assume that ε is normally distributed with zero mean and finite variance σ^2 we can (1) relate the observed mean square deviation to σ^2, and (2) determine the probability that a particular observation falls within a given interval around the predicted value.

If we define L as

$$L = \frac{1}{N} \sum_n d_n^2 = \frac{1}{N} \sum_n [y_n - \hat{y}_n]^2 = \langle [y - \hat{y}]^2 \rangle$$

$$= \frac{1}{N} \sum_n [y_n - \beta x_n - \alpha]^2 = \frac{1}{N} \sum_n [\varepsilon_n]^2.$$

$L = \sigma^2$, and as is shown in the Appendix,

$$L = \langle \delta y^2 \rangle - \frac{\langle \delta x \delta y \rangle^2}{\langle \delta x^2 \rangle} = \sigma^2.$$

The statistic d will be normally distributed because, by assumption, the error term ε is normally distributed with $z = d/\sigma$. Thus, because the normal distribution is

$$p(z) = \frac{1}{\sqrt{2\pi}} \exp\left(-\frac{z^2}{2}\right),$$

the probability $P(z_c)$ that an observation will lie between $-z_c$ and $+z_c$ is given by :

$$P(z_c) = \int_{-z_c}^{z_c} p(z) dz = \int_{-z_c}^{z_c} \frac{1}{\sqrt{2\pi}} \exp\left(-\frac{z^2}{2}\right) dz.$$

So, we now have a procedure for determining a confidence interval for the deviation between the observation and the predicted value of y.

First, one specifies the degree of confidence one seeks; for example, 95% confidence that $z = d/\sigma$ will lie between $-d_c/\sigma$ and $+d_c/\sigma$.

Next, one looks up in a table (or computes numerically) the value z_c that gives 0.95 of the area under the normal curve. For example,

Confidence level(%)	50	90	95	99
z_c	0.6745	1.645	1.96	2.58

The result is a predicted curve with upper and lower bounds that define the confidence interval:

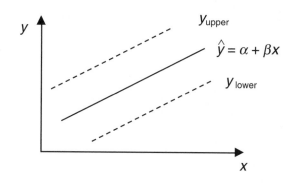

where

$$y_{\text{upper}} = \alpha + \beta x + d_c$$
$$\hat{y} = \alpha + \beta x$$
$$y_{\text{lower}} = \alpha + \beta x - d_c$$

ESTIMATING RECENT TRENDS IN ENERGY EFFICIENCY IN THE U.S.

We next apply these tools to an analysis of recent trends in U.S. energy consumption. The data we use are taken from the EIA web page on the Department of Energy web site. The data are drawn from the EIA Annual Energy Review (AER) data base and are shown in Table 14.11.

Example 1. Is Recent Energy Use Correlated with Real GDP?
As our first example, we inquire how closely recent energy use is correlated with GDP. We wish to determine whether the energy intensity of the economy is declining. To examine this proposition we assume a linear relationship between the dependent variable $y = E/1990$, the normalized energy use in the table, and a single independent variable $x = X/1990$, normalized real GDP.

From the data in the table we find:

$$\langle y \rangle = 1.0497 \qquad \langle x \rangle = 1.0646 \qquad \langle \delta y^2 \rangle = 0.00255 \qquad \langle \delta x^2 \rangle = 0.00671$$

Table 14.11. U.S. energy trends, 1988–98

Year	Real GDP, X (billion $)	Energy consumption, E (quad)	Real electricity price, P (industrial) (mills/kwh)	Heating degree-days, HDDs	Normalized energy use (E/1990)	Normalized GDP (X/1990)	Normalized HDDs (HDD/1990)	Normalized electricity price (P/1990)
1988	5,865.20	83.04	55	4,653	0.9872	0.9558	1.1586	1.1
1989	6,062.00	84.53	52	4,726	1.0049	0.9879	1.1768	1.04
1990	6,136.30	84.12	50	4,016	1	1	1	1
1991	6,079.40	84.03	49	4,200	0.9989	0.9907	1.0458	0.98
1992	6,244.40	85.49	48	4,441	1.0163	1.0176	1.1058	0.96
1993	6,389.60	87.31	47	4,700	1.0379	1.0413	1.1703	0.94
1994	6,610.70	89.26	46	4,483	1.0611	1.0773	1.1163	0.92
1995	6,761.70	91	44	4,531	1.0818	1.1019	1.1282	0.88
1996	6,994.80	93.97	42	4,713	1.1171	1.1399	1.1736	0.84
1997	7,269.80	94.37	40	4,542	1.1218	1.1847	1.131	0.8
1998	7,552.10	94.23	40	4,029	1.1202	1.2307	1.0032	0.8
AVERAGE					1.049745	1.06456	1.109964	0.932727
VARIANCE					0.002554	0.00671	0.003906	0.008529

Source: U.S. Energy Information Administration

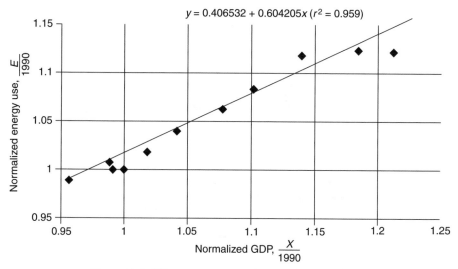

Figure 14.5. U.S. energy consumption vs. GDP correlation.

The covariance is: $\langle \delta y \delta x \rangle = 0.00405$.

Using the previous formulas it is straightforward to compute the relevant statistical quantities:

$$\beta = \frac{\langle \delta y \delta x \rangle}{\langle \delta x^2 \rangle} = 0.6036 \qquad \alpha = 0.407132$$

Thus, $y = 0.4071 + 0.6036x$ is the least squares fit. The result of the correlation with data points is plotted in Figure 14.5. (The Excel program finds slightly different parameters, as noted on the figure.)

We can easily calculate the correlation coefficient from the formula:

$$r^2 = \frac{\langle \delta x \delta y \rangle^2}{\langle \delta x^2 \rangle \langle \delta y^2 \rangle} = 0.959$$

which is a high correlation.

Let us now find the confidence intervals for this correlation. We use the formula

$$\sigma^2 = \langle \delta y^2 \rangle - \frac{\langle \delta x \delta y \rangle^2}{\langle \delta x^2 \rangle} = \langle \delta y^2 \rangle [1 - r^2] = 0.00255\,[1 - 0.9591] = .000104.$$

Therefore, $\sigma = 0.010212$. Since $z = d/\sigma$ and $z = 1.96$ at the 95% confidence interval, we have $d = 0.02$. Thus, a crude estimate of the 95% confidence interval is

$$y_{\text{upper}} = 0.4271 + 0.6036x$$
$$y = 0.4071 + 0.6036x$$
$$y_{\text{lower}} = 0.3871 + 0.6036x.$$

The difference between observed and predicted energy use, referred to as the "residual,"

Table 14.12. Observed and predicted U.S.
energy consumption, 1988–98

Year	Observed E	Predicted E	Residual
1988	0.9872	0.98403106	0.00316894
1989	1.0049	1.00342603	0.00147397
1990	1	1.0107369	−0.0107369
1991	0.9989	1.0051178	−0.0062178
1992	1.0163	1.02137091	−0.0050709
1993	1.0379	1.03569056	0.00220944
1994	1.0611	1.05744192	0.00365808
1995	1.0818	1.07230536	0.00949464
1996	1.1171	1.09526513	0.02183487
1997	1.1218	1.1223335	−0.0005335
1998	1.1202	1.13948083	−0.0192808

is given in Table 14.12 (using the Excel parameters). Note that during the two most recent years the deviation is negative and growing. We shall come back to this observation in Example 3 below.

Example 2. Are Energy Use and Energy Price Correlated?
As a second example, we consider whether the recent trends in energy use and energy prices are correlated. Here the question is whether we can attribute the increase in energy use to lower energy prices. (The expectation is that lower prices will cause businesses and consumers to use more energy.)

For the sake of simplicity, we take one measure of energy prices – the wholesale price of electricity in the United States – and correlate this with energy use over the last eleven years. Electricity prices reflect, among other factors, the price of the fuel used to produce the electricity. In the United States, the fuel mix importantly includes coal, natural gas, and nuclear.

The results of the correlation are summarized in Figure 14.6. The correlation is not as strong as in Example 1 but with $r^2 = 0.91$ it is still quite good. Over the eleven year period energy use increased and energy prices declined.

From the correlation we are tempted to conclude that there is a causal relationship between increasing energy use and decreasing energy prices. But we must remember that the reported data do not represent either supply or demand curves, but rather the market clearing intersection of supply and demand. While both theory and common sense lead us to expect that demand will increase as price declines, theory does not require that the market clearing price should necessarily either increase or decrease as the market clearing quantity increases. This is illustrated by the two situations in Figure 14.7.

In the first case (panel (a)), the supply curve is assumed to shift over time, so that a greater quantity of energy is offered at each price at the later time. In this case we see that the market clearing price decreases from p to p' as the market clearing quantity increases from q to q'. This seems logical and fits the correlation we have found. In

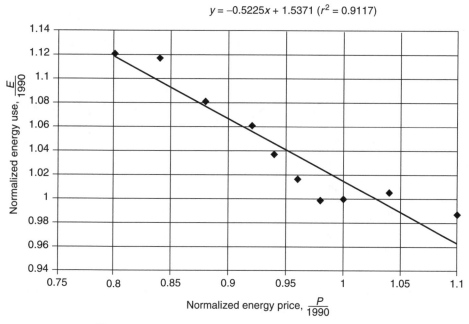

Figure 14.6. U.S. energy consumption vs. price correlation.

the second case, however, the supply curve is unchanged while the demand curve is assumed to move with time so that at the later time there is more demand at any given price. In this case we see (from panel (b)) that the market clearing price increases from p to p', while the market clearing quantity also increases from q to q'. The point is that theory does not tell us what the correlation between market clearing prices and quantities should be.

There are many reasons why an observed correlation between two variables may not establish a causal relationship. The correlation may simply reflect two independent

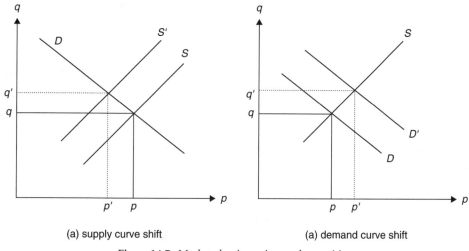

Figure 14.7. Market clearing prices and quantities.

trends. There may be additional variables that influence either variable that have not been included in the assumed regression. Finally, there may be a time lag between the response of one variable to a change in the other. This example illustrates the limitations of the power of statistical inference.

Example 3. Energy Use and GDP, Including the Effect of Weather
As previously noted, the correlation between energy use and GDP showed an increasing (negative) deviation between actual and predicted energy values in 1997 and 1998. During those two years, there was a sizeable increase in economic output but hardly any increase in energy use.

Year	Normalized energy use $E/1990$	Normalized GDP $X/1990$
1996	1.1171	1.1399
1997	1.1218	1.1847
1998	1.1202	1.2307

Does this indicate a welcome shift from the historical pattern of energy use and economic activity examined in Example 1? Or is the reported data merely a statistical fluctuation that should be expected and signifies nothing? The answer is quite important. If the trend is significant, it is very good news for those concerned with improving energy efficiency and avoiding global warming. If the trend is not significant, it suggests that the reported statistics have sufficient "noise" to conceal important indicators of change. This conclusion might lead to making a greater effort to collect accurate energy statistics.

Using the correlation equation $E = 0.604205X + 0.406532$ calculated by the Excel program, the regressions for 1997 and 1998 are as follows:

Year	$E/1990$ predicted	$E/1990$ reported	Residual
1997	1.122334	1.1218	−0.0005
1998	1.139481	1.1202	−0.01928

The 1998 residual is large and negative. (It is not the largest residual of the 11 points in the correlation. The 1996 residual is larger, +0.022, but it does not attract the same attention because it goes in the "wrong" direction from the viewpoint of improving energy efficiency.)

The first question that a policy analyst should ask is, "Does the 1998 observation lie outside the 95% confidence interval?" For 1998, $X = 1.21308$ and the 95% upper and lower confidence intervals are: $E_{upper} = 1.15944$ and $E_{lower} = 1.11943$. The 1998 observed value, 1.1202, lies within this confidence interval and thus should not be taken as evidence of a startling deviation that might indicate a shift towards a pattern of more efficient energy use.

Nevertheless, the 1998 observation has a large residual. Is there some other variable that might explain this deviation? Statisticians continually confront this question. No matter how good the correlation between a predicted value and an observation might be, there is always the desire to include some other variable that "explains" – that is, reduces – the remaining residual.

Table 14.13. Observed and predicted
U.S. energy consumption, 1988–98
(two-variable regression)

Year	Predicted	Reported	Residual
1988	0.989029	0.9872	−0.001823
1989	1.01129	1.0049	−0.006383
1990	0.995206	1	0.004791
1991	0.995563	0.9989	0.003336
1992	1.02018	1.0163	−0.003876
1993	1.04341	1.0379	−0.005510
1994	1.05846	1.0611	0.002640
1995	1.07524	1.0818	0.006556
1996	1.10477	1.1171	0.012331
1997	1.12677	1.1218	−0.004974
1998	1.12729	1.1202	−0.007085

In this particular case there is another variable that might be expected to play an important role – the weather. If the weather is unseasonably hot or cold, this will influence energy consumption. In fact, 1998 was an uncommonly warm year, so this should have been reflected in reported energy use. A particularly motivated policy analyst might ask for a regression to be run for two variables: GDP and a variable that measures the effect of weather. The EIA uses "Heating Degree Days" as a measure of how cold a region was over a period of time relative to a base temperature (usually taken to be 65°F). The number of heating degree days (HDD) for a single day in a region is the difference between the reference temperature and the day's average temperature, if the difference is positive. If the difference is negative, and the average day's temperature T exceeds the reference temperature, T_{ref}, then the HDD value is zero. Thus,

$$HDD = Max\{(T_{ref} - T), 0\}.$$

Cumulative HDD measures the number of degree-days that fall below the reference temperature and, by implication, require some heat and thus energy use. Cold years have high HDDs and warm years have low HDDs. (The EIA HDD data are averaged over a number of climate regions in the United States).

The next step is to carry out a two-variable regression, predicting E/1990, on the basis of GDP/1990 and HDDs/1990. The procedure that must be employed is presented in the Appendix. There are many computer programs that will carry out such two (or more) variable regressions quite rapidly. We used the regression program that is included in the 1998 Microsoft Office Excel program. For the 1988–1998 data displayed in Table 14.11, we obtain the least squares result:

$$(E/1990) = 0.244156 + 0.617819 \, (GDP/1990) + 0.133231(HDD/1990).$$

The resulting predicted and reported values of E/1990 along with the residuals are presented in Table 14.13. Including the effect of weather has "explained" two thirds of the 1998 residual. Using only the GDP as an explanatory variable led to a residual of −0.019; using both the GDP and weather reduced the residual to 0.007.

This example illustrates the procedure and power of statistics. Once a correlation is established, the analyst is naturally drawn to explore additional variables that lead to an improved correlation, that is, a reduction in residuals. While this process is referred to as "explaining" the data, it is important to keep in mind that no correlation, however good, proves causality. This example is of importance in its own right, because it is a real case in which policymakers were eager to use one year of data to make a political point about a trend of improving energy efficiency in the country. Only the intervention of determined analysts prevented the policymakers from making a mountain out of a molehill of no statistical significance. Indeed, the variability in reported energy statistics means that it will take considerable time to confirm any trend in improving energy efficiency in the U.S. economy.

APPENDIX TO CHAPTER 14

In this appendix, we present an introductory treatment for determining coefficients in linear regressions based on minimization of the sum of the deviations between predicted and observed values.

The basic rule is to choose the coefficients in such a way that the average of the sum of the squares of the deviations between the actual observations and the predicted values is minimized. Thus, we define the deviation $d_n = y_n - \hat{y}_n$, and

$$L = \frac{1}{N} \sum_n d_n^2 = \frac{1}{N} \sum_n [y_n - \hat{y}_n]^2 = \langle [y - \hat{y}]^2 \rangle = \frac{1}{N} \sum_n [y_n - \beta x_n - \alpha]^2.$$

It is convenient to express the quantities in terms of deviations from their means:

$$\delta F_n = F_n - \langle F \rangle,$$

so that L can be expressed as:

$$L = \frac{1}{N} \sum_n [(\delta y_n - \beta \delta x_n) + (\langle y \rangle - \beta \langle x \rangle - \alpha)]^2,$$

which works out to be:

$$L = \langle \delta y^2 \rangle - 2\beta \langle \delta y \delta x \rangle + \beta^2 \langle \delta x^2 \rangle + [\langle y \rangle - \alpha - \beta \langle x \rangle]^2.$$

We determine the coefficients (α, β) by minimizing L:

$$\frac{\partial L}{\partial \alpha} = -2[\langle y \rangle - \alpha - \beta \langle x \rangle] = 0,$$

so that: $\langle y \rangle = \beta \langle x \rangle + \alpha$; *Note:* $\langle y \rangle = \langle \hat{y} \rangle$.

The condition for β comes from:

$$\frac{\partial L}{\partial \beta} = -2 \langle \delta y \delta x \rangle + 2\beta \langle \delta x^2 \rangle - 2 [\langle y \rangle - \alpha - \beta \langle x \rangle] \langle x \rangle = 0,$$

so that

$$\beta = \frac{\langle \delta y \delta x \rangle}{\langle \delta x^2 \rangle}.$$

(We do not prove here that this stationary point corresponds to a minimum.)

At the stationary point, the value of L is just

$$L = \langle \delta y^2 \rangle - 2\beta \langle \delta y \delta x \rangle + \beta^2 \langle \delta x^2 \rangle = \langle (\delta y - \beta \delta x)^2 \rangle = \left\langle \left(\delta y - \frac{\langle \delta y \delta x \rangle}{\langle \delta x^2 \rangle} \delta x \right)^2 \right\rangle.$$

This can be expressed in a number of alternative forms. A particularly useful expression is:

$$\frac{L}{\langle \delta y^2 \rangle} = \left[1 - \frac{\langle \delta y \delta x \rangle^2}{\langle \delta y^2 \rangle \langle \delta x^2 \rangle} \right].$$

The "correlation coefficient," r, is defined so as to give a measure of the "goodness of fit" of the approximate form. The definition of "r" is:

$$r^2 = \frac{\text{Predicted Variation}}{\text{Total Variation}} = \frac{\langle \delta \hat{y}^2 \rangle}{\langle \delta y^2 \rangle}.$$

We have $\langle \delta y^2 \rangle = \langle [\delta \hat{y} + (y - \hat{y})]^2 \rangle = \langle \delta \hat{y}^2 \rangle + 2 \langle \delta \hat{y} (y - \hat{y}) \rangle + L$.

The cross term can be expressed as:

$$\langle \delta \hat{y} (y - \hat{y}) \rangle = \langle \delta \hat{y} (\delta y - \delta \hat{y}) \rangle = \beta \langle \delta x \delta y \rangle - \beta^2 \langle \delta x^2 \rangle,$$

which vanishes when one of the factors of β in the last term is replaced by the value determined at the minimum. Thus,

$$r^2 = \frac{\langle \delta \hat{y}^2 \rangle}{\langle \delta y^2 \rangle} = 1 - \frac{L}{\langle \delta y^2 \rangle} = \frac{\langle \delta y \delta x \rangle^2}{\langle \delta y^2 \rangle \langle \delta x^2 \rangle}.$$

If there is no correlation between y and x, then $r = 0$ and $L = \langle \delta y^2 \rangle$; if there is perfect correlation, then $r = 1$, and $L = 0$. Note that $0 \leq r \leq 1$.

Generalization to Several Variables

It is not a very difficult task to generalize the treatment to the case of s independent variables. In this case

$$y = \alpha + \beta_1 x_1 + \cdots + \beta_s x_s + \varepsilon,$$

which in vector notation is

$$y = \alpha + \vec{\beta}^T \cdot \vec{x} = \alpha + \vec{x}^T \cdot \vec{\beta} + \varepsilon.$$

Here the dimension of each vector is the number of variables. The predicted relation is of the form: $\hat{y} = \alpha + \vec{\beta}^T \cdot \vec{x}$, so

$$L = \left\langle [y - \hat{y}]^2 \right\rangle = \left\langle [\delta y - (\hat{y} - \langle y \rangle)]^2 \right\rangle$$

$$L = \left\langle [(\delta y - \vec{\beta}^T \cdot \delta \vec{x}) - (\alpha + \vec{\beta}^T \cdot \langle \vec{x} \rangle - \langle y \rangle)]^2 \right\rangle$$

$$L = \left\langle (\delta y - \vec{\beta}^T \cdot \delta \vec{x})^2 \right\rangle + \left\langle (\alpha + \vec{\beta}^T \cdot \langle \vec{x} \rangle - \langle y \rangle)^2 \right\rangle$$

$$= \left\langle (\delta y - \vec{\beta}^T \cdot \delta \vec{x})^2 \right\rangle + (\alpha + \vec{\beta}^T \cdot \langle \vec{x} \rangle - \langle y \rangle)^2.$$

We now minimize L by setting the first derivatives with respect to α and each component of $\vec{\beta}$ equal to zero

$$\frac{\partial L}{\partial \alpha} = -2(\alpha + \vec{\beta}^T \cdot \langle \vec{x} \rangle - \langle y \rangle) = 0 \text{ , therefore } \langle y \rangle = \alpha + \vec{\beta}^T \cdot \langle \vec{x} \rangle .$$

The derivatives with respect to the β_i must satisfy:

$$\nabla_{\vec{\beta}} L = -2\langle [\delta y - \vec{\beta}^T \cdot \delta \vec{x}] \delta \vec{x} \rangle - 2[\alpha + \vec{\beta}^T \cdot \langle \vec{x} \rangle - \langle y \rangle] \langle \vec{x} \rangle = 0.$$

These equations are satisfied (taking into account the condition for the α derivative above) by the relations:

$$\langle [\delta y - \vec{\beta}^T \cdot \delta \vec{x}] \delta \vec{x} \rangle = 0 \text{ , or } \vec{\beta}^T \cdot \langle \delta \vec{x} \delta \vec{x} \rangle = \langle \delta y \delta \vec{x} \rangle .$$

Since

$$\vec{\beta}^T \cdot \langle \delta \vec{x} \delta \vec{x} \rangle = \langle \delta \vec{x} \delta \vec{x} \rangle \cdot \vec{\beta},$$

the result is:

$$\vec{\beta} = \langle \delta \vec{x} \delta \vec{x} \rangle^{-1} \cdot \langle \delta y \delta \vec{x} \rangle ,$$

where $\langle \delta \vec{x} \delta \vec{x} \rangle^{-1}$ is the inverse matrix of the matrix $\langle \delta \vec{x} \delta \vec{x} \rangle$. Thus, $\langle \delta \vec{x} \delta \vec{x} \rangle^{-1} \cdot \langle \delta \vec{x} \delta \vec{x} \rangle = \vec{\vec{I}}$, where $\vec{\vec{I}}$ is the identity matrix.

Example for Two Variables

Let us assume the linear regression $y = \alpha + \beta_1 x_1 + \beta_2 x_2 + \varepsilon$. Minimizing the sum of the square of the deviations leads to the results:

$$\langle y \rangle = \alpha + \beta_1 \langle x_1 \rangle + \beta_2 \langle x_2 \rangle ,$$

and for the values of the βs:

$$\begin{bmatrix} \beta_1 \\ \beta_2 \end{bmatrix} = \begin{bmatrix} \langle \delta x_1 \delta x_1 \rangle & \langle \delta x_1 \delta x_2 \rangle \\ \langle \delta x_2 \delta x_1 \rangle & \langle \delta x_2 \delta x_2 \rangle \end{bmatrix}^{-1} \cdot \begin{bmatrix} \langle \delta y \delta x_1 \rangle \\ \langle \delta y \delta x_2 \rangle \end{bmatrix}.$$

For a 2×2 matrix the inverse is easily computed:

$$\begin{bmatrix} \langle \delta x_1 \delta x_1 \rangle & \langle \delta x_1 \delta x_2 \rangle \\ \langle \delta x_2 \delta x_1 \rangle & \langle \delta x_1 \delta x_1 \rangle \end{bmatrix}^{-1} = \frac{1}{\langle \delta x_1 \delta x_1 \rangle \langle \delta x_1 \delta x_1 \rangle - \langle \delta x_1 \delta x_2 \rangle^2} \begin{bmatrix} \langle \delta x_1 \delta x_1 \rangle & -\langle \delta x_2 \delta x_1 \rangle \\ -\langle \delta x_1 \delta x_2 \rangle & \langle \delta x_2 \delta x_2 \rangle \end{bmatrix}.$$

These formulas are used for the statistical regression analysis of Example 3.

15

The Government's Role in Innovation

The fundamental purpose of national government is to protect and improve the public welfare. Governments seek to defend national security (in the broadest sense), work to improve the economic well-being of their citizens, and protect public health, safety, and natural resources. Technological advance plays a vital role in each of these domains. The government, therefore, has a strong interest in encouraging technological innovation. But exactly what should this role be? How can the government be involved most effectively in the development and implementation of new technologies? What policy instruments can and should the government use to intervene in the innovation process?

The pace and direction of innovation depends on the research and development (R&D) activity of the economy. As Figure 15.1 shows, each year U.S. industry and government both devote large amounts of resources to R&D.

The R&D activity reported in these statistics is only part of the innovation process, of course. As we have seen, inventing and developing technology is just one of the steps involved. Invention must be followed by implementation, which is a complex, uncertain, and costly process. Our focus in this chapter is primarily on R&D, especially the R&D that is funded by the U.S. government at a current rate of roughly $75 billion per year. By any measure, this is a lot of money. But is it too much? Or not enough? As Figure 15.2 shows, the federal government's share of total U.S. R&D expenditures has declined from a peak of about two-thirds in the early 1960s to about 25% of the total today. Again we might ask: Is this too low? Too high? Just right?

There is no precise formula we can invoke to answer such questions. Partly this is because R&D, like any investment, involves a trade-off between current and future consumption – a matter of social preference for which there is no "right" answer. But another problem that is unique to R&D is the great uncertainty in the character and timing of the benefits that any particular R&D project or program will generate. This uncertainty is often used to justify the use of public monies – the government can take risks that industry cannot or will not take – but it simultaneously makes the justification process much more complicated.

To address these problems we begin by briefly describing both the character of the R&D enterprise and its relationship to the implementation of new technologies.

A Taxonomy of R&D and Innovation

The usual way of describing R&D is to position the various kinds of R&D activity along a single axis. At one end of this axis is "pure" research – fundamental in nature,

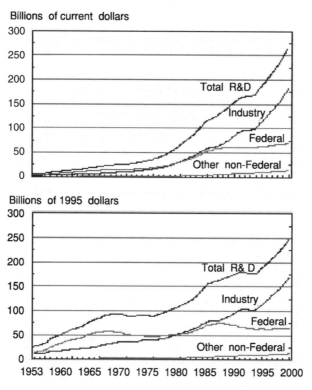

Source: Reprinted from National Science Board, *Science and Engineering Indicators – 2002.*

Figure 15.1. U.S. R&D funding by source, 1953–2000.

and driven by the curiosity of the individual investigator. At the other end is "applied" work that is motivated by the practical desire or need to develop a useful product or service.

This classification of R&D corresponds to a particular view of innovation as a process that flows linearly through time. In this view, innovation starts with basic research and then proceeds sequentially through development to implementation, with different skills involved at each stage and without much emphasis on communication between the stages. The linear innovation model is pictured as:

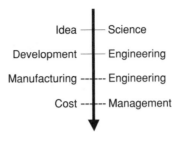

This model is well entrenched. For example, the Department of Defense manages R&D according to rigid budget categories, 6.1 through 6.5, that map directly onto it:

6.1 basic research
6.2 exploratory research
6.3 advanced development
6.4 engineering development
6.5 system development

The linear model of innovation is also reflected in the structure of U.S. higher education. By organizing their educational programs into schools of science, engineering, and management, universities implicitly endorse the linear model.

The linear model is not wrong per se. There are circumstances in which it properly characterizes reality. For much of the thirty-year period from 1960 to 1990, for example, the United States made tremendous technical advances by applying modern science, especially physics, to the development of new systems with unprecedented performance characteristics such as satellites, jet aircraft, and computers. Many of these advances came out of the effort the United States made to acquire more capable military systems than those of the Soviet Union, our Cold War adversary. Superior performance was the priority, not cost, and the government, not the commercial marketplace, was the most important customer for high technology. During this period, major U.S. corporations such as IBM, AT&T, and DuPont were also vigorously exploiting the scientific knowledge generated by their big central research laboratories, and they too were reaping the rewards of applying science to technology development. Creating new ideas was regarded as the main task of innovation; implementation was thought to be relatively easy.

By the late 1980s, it was becoming clear that something was wrong with this model of innovation. Industry in the United States seemed chronically unable to improve its productivity, while Japanese firms, which had made major strides in both productivity and quality, were capturing growing shares of U.S. and world markets for autos, consumer electronics, semiconductors, and other products. Many industry and academic groups became concerned about the competitiveness of U.S. industry during this period and called for government assistance programs that would improve the nation's productivity and innovation performance.

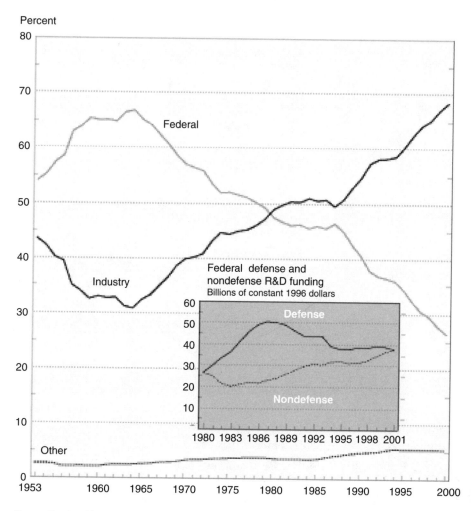

Source: Reprinted from National Science Board, *Science and Engineering Indicators — 2002*.

Figure 15.2. Industry and government shares of U.S. R&D funding, 1953–2000.

This was a major shift in government technology policy priorities – from advancing national security to promoting the well-being of the civilian economy. The change in focus was made possible, of course, by the collapse of the Soviet Union and the end of the Cold War. The new focus prompted a re-examination of how R&D should be organized so as to advance economic growth. It led to greater emphasis in policy on "downstream" development activities that were closer to the point of technology adoption in the marketplace. But concerns were raised. Some believed that the R&D process, especially basic research, would be compromised by the attempt to make connections to specific economic ends. Others opposed any government role in "industrial policy," on the grounds that the government could not be successful at picking technology winners for the private sector.

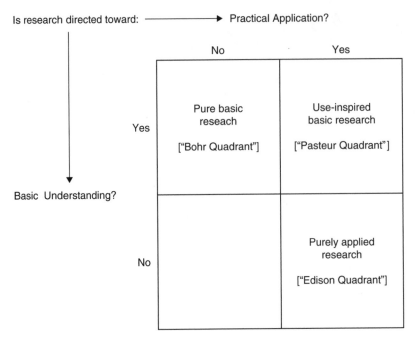

Figure 15.3. Stokes' research and development matrix.

Donald Stokes, who was Dean of the Woodrow Wilson School at Princeton University, presented an important conceptual analysis which pointed out that the goals of fundamental research and practical application – the two poles of the linear model – are not mutually exclusive.[1] As Stokes observed, some research that is "basic" in character, meaning that that it produces new knowledge about fundamental scientific phenomena, may nevertheless be driven by very practical goals. A good example is research on the structure and function of the human genome. It would be difficult to think of anything more "fundamental" than this, yet such work is driven by the practical goal of improving human health (and for some researchers also by the prospect of financial reward).

Stokes suggested that R&D activity could be better understood if it was mapped onto a two-dimensional plot rather than on a line. In Stokes' plot, one axis describes the broad purpose of the research: Is the new knowledge being pursued entirely for its own sake, or rather with some practical considerations of use in mind? The other has to do with the character of the research: Does it advance the frontiers of fundamental scientific knowledge or not? Stokes used these distinctions to create the simple two-by-two matrix in Figure 15.3.

The upper-left cell refers to fundamental research that is pursued purely for its own sake with no thought of practical application. A good example is the research on the structure of the atom carried out early in the twentieth century by the great Danish

---

[1] Donald E. Stokes, *Pasteur's Quadrant: Basic Science and Technological Innovation*, Brookings Institution, Washington, D.C., 1997.

physicist Niels Bohr. This was a pure voyage of discovery at the time (although later, of course, it would have enormous practical consequences, as the world would discover at Hiroshima and Nagasaki). Stokes called this cell the "Bohr quadrant."

The lower-right quadrant includes research that is pursued solely with practical goals in mind and with no thought of advancing the understanding of a scientific field. This is epitomized by the work of Thomas Edison, who took existing knowledge and converted it into practical inventions while paying almost no attention to the deeper scientific implications of his discoveries. Much industrial R&D today falls into this category. Stokes called it the Edison quadrant.

Then there is the upper-right cell: research at a fundamental level that is motivated by practical concerns. We have mentioned human genome research already. A much earlier example was the research of Louis Pasteur (1822–1895), whose very practical goal of treating specific diseases in humans and animals and preventing the spoilage of milk and other foods led him to undertake fundamental investigations of the microbiological foundations of disease. Stokes called this cell the Pasteur quadrant. (Work in the lower-left quadrant neither seeks to explain at a fundamental level nor has any practical purpose in mind. Stokes cited Roger Tory Peterson's taxonomic studies of birds. But it is not a significant category for our purposes, and we need not consider it further.)

The Pasteur quadrant is of particular interest because it describes an important category of R&D activity that is fundamental in nature yet motivated by practical goals. This type of R&D potentially can uncover knowledge that has implications for many private firms. For a single firm to hold such knowledge proprietary would thus not be desirable from a public policy perspective. Advancing such "pre-competitive" technology that could improve the productivity and innovation performance of many firms became an important objective of government R&D programs in the early 1990s. We return to this topic later in the chapter.

At the same time that government technology policy priorities were shifting, a parallel change was taking place in the understanding of the innovation process. Most applications of technology in the civilian economy, if they are to succeed, require attributes that are quite different from military technologies. In place of technical performance (the driving force of the linear model), relevant attributes are quality, affordable cost and, importantly, time to market. In order to achieve these desiderata, a very different framework for understanding innovation has emerged that stresses the parallel, interactive nature of successful innovation. In contrast to the linear model, in the new innovation framework all of the factors necessary for successful commercial innovation must take place almost simultaneously, and in an integrated way. R&D, manufacturing, cost, and regulations are considered from the outset of the development process. For example, design for manufacturability begins early in the development process because this reduces cost and development lead-time. System integration is stressed as much as component development. An integrated product development team that can bring together all of the necessary skills carries out the R&D and innovation process.

Clearly this integrated approach suggests that the barriers to successful innovation lie as much in implementation as in the generation of new ideas. If useful knowledge can be generated anywhere in the value chain, including the firm's customers and suppliers, what becomes critical is the ability to acquire this knowledge (i.e., to learn) and to bring to bear the full range of capabilities – in R&D, manufacturing, engineering development, marketing, design, and so on – that are needed to convert the knowledge into something of value to the firm's customers as quickly as possible. This calls for a flexible organization, for closer links with customers and suppliers, and for a research effort that is integrated as fully as possible into the business enterprise. Many fundamental researchers have adopted the Pasteur quadrant view and broadened their interest from simply making research advances to include an interest in the application of new ideas.

The success of this parallel innovation paradigm is also heavily dependent on advances in computers, communications, and analytic modeling and simulation techniques that permit groups to work together with computer-aided engineering, design, and manufacturing tools.

In recent years, U.S. industry has aggressively adopted these organizational and technological approaches to innovation, and most knowledgeable observers agree that this contributed significantly to the almost ten years of uninterrupted economic growth enjoyed by the United States during the 1990s. Technological innovation has proceeded so rapidly in the civilian sector, especially in the important field of information technology, that government is no longer the source of the most advanced technology, except for defense-specific items such as nuclear submarines. But this new model of innovation also makes the role of the federal government in R&D more complicated because the two processes of generating new knowledge and of implementing this knowledge are harder to distinguish from each other, both conceptually and in practice.

We can now consider several important questions. When should the government intervene in support of the innovation process? In cases where government intervention is warranted, what kind of intervention is called for? When should the government pay directly for R&D? What other policy instruments are available, and when should they be used? When the government does intervene, who should make the decisions about what specific scientific or technological objectives to pursue? When the government pays directly for R&D, who should own the intellectual property that is created? And who should actually perform the R&D: Universities, other non-profit research institutions (e.g., hospitals), private corporations, or national laboratories? The answers to these questions depend, of course, on the purpose of the government's support.

RATIONALES FOR GOVERNMENT INTERVENTION IN SUPPORT OF INNOVATION

There are several different circumstances in which a government role in support of the innovation process can be justified. First, the government clearly has such a role in areas where it is itself the primary or exclusive user of specialized technologies needed to conduct the public's business. The most prominent example is that of national defense. Other examples include law enforcement and tax collection. As we have seen, government support for R&D bearing on national security was the dominant force in technology advance and innovation during the Cold War period. Not only did it give the United States a technological advantage over the Soviet adversary in a broad range of military areas, from submarines to combat aircraft to battlefield communication, but it also had a huge, unanticipated payoff in the civilian economy. Technologies supported by agencies within the Department of Defense, notably the Office of Naval Research and the Defense Advanced Research Projects Agency (DARPA), which was founded in 1958 in response to the Soviet launching of the first Sputnik satellite, gave birth to vast new civilian industries and led directly to the information technology revolution that has now affected the life of every citizen.

National defense and, on a smaller scale, law enforcement and tax collection are concerns exclusive to the government. In other cases, the government plays a role or provides a service that is a necessary adjunct to private sector activity – for example, the federal air traffic control system, or, as we have seen in an earlier chapter, the disposal of high level nuclear waste, or the federal interstate highway system, or the allocation of the electromagnetic spectrum for communications. In still other cases, the private sector is the primary service provider, but there are important common or collective goods involved for which the government assumes responsibility. The best example is that of the health care system. In all these areas, technological advance is important to the government's ability to improve its performance, and there is, thus, a clear justification for government support of the innovation process.

A fourth domain in which government support is warranted concerns the fundamental discoveries that underlie much innovative activity. Examples abound in the fields of physics and chemistry, material science and engineering, mathematics, space science, linguistics, and so on. Much of this kind of research lies squarely in Stokes' Bohr quadrant. All agree that such knowledge may be of great value to society in the future, but it is impossible to predict when or how. To increase the likelihood that these uncertain future benefits will be captured, the research results should be unrestricted and made freely available to all. Accordingly, it is unreasonable to expect private firms, acting either alone or in concert, to bear the cost of this type of research. So it makes sense for "Bohr"-type research to be financed by the government, and to publish the results in the open literature. The government has other reasons to fund this kind of research too. One is the belief that society is culturally and intellectually enriched by scientific discovery. Another, more practical motivation is that fundamental research, much of which is carried out at universities, simultaneously serves to train the next generation of scientists and engineers.

Not all federally funded fundamental research is of the Bohr type. Much of it, in fact, is conducted in support of one or another of the government's missions – for example, national defense or public health – and so more properly belongs in the Pasteur quadrant. As long as there is a clear connection to a public mission the case for government support is clear. More problematic, however, is Pasteur-type fundamental research that is aimed at strengthening the technological base of that part of the economy where the government does not have a direct role. This kind of research is typically relevant not just to a specific product but to a broad class of products or services, and therefore is likely to be of interest to a broad range of firms. Still, if private firms are the direct beneficiaries of such R&D, why should the taxpayer pay for it?

Most of the time, in fact, there is no case for public funding. This is true of most of the R&D in the Edison quadrant, for example. But situations can arise, even in the Edison quadrant and more frequently in the Pasteur quadrant, in which private firms will systematically underinvest in R&D relative to the socially optimal level because they are not able to capture the full benefits of the research themselves. In such situations a good case can be made for government support. But when do such situations arise? Identifying them is not always easy. The fact that a private firm is not pursuing a particular line of R&D may simply mean that it is not a good business investment. Private firms will generally invest their own funds in R&D on new products and processes if the potential returns on these investments are large enough to offset the technological and market risks involved. Just because a firm chooses not to proceed with a particular R&D project that some people think is attractive does not automatically mean that public funding is needed instead.

But there are some situations in which private investment calculations are negatively affected by market imperfections or failures, and here the case for government intervention becomes stronger. If, for example, the price of products or services in a particular market sector is held artificially low by government price regulations, private firms will have fewer incentives to invest their own funds in R&D (or, for that matter, in any other type of capital-seeking activity). In the late 1970s, when the price of domestic oil was still regulated, U.S. industry had very little incentive to undertake R&D aimed at improving domestic oil exploration or production technologies. To compensate, the Department of Energy granted oil companies that undertook R&D projects on enhanced oil recovery (EOR) technologies (and published their results) an entitlement to receive the higher world price for their EOR production.

Private investment will also be reduced if firms face trade restrictions that prevent or hinder them from selling their products into particular country markets, or if foreign governments subsidize their firms' exports to the United States. During the late 1980s, many proposals were put forward to provide R&D subsidies to American firms in order to offset what were perceived to be unfair subsidies that the Japanese government was giving to Japanese companies who were exporting to U.S. markets.

Another type of market imperfection that might justify government intervention has to do with external costs or benefits. If the full benefit or cost of an economic activity is not properly reflected in market prices, then private firms will tend to underinvest in innovation. For example, if the environmental costs of acid rain or greenhouse gas emissions are not fully reflected in the market price of fossil fuels, private firms will invest too little in energy conservation or renewable energy technologies from a social point of view. As another example, the vulnerability of the United States and its allies to world oil market disruptions and the cost of defending against the risk of such disruptions implies that a security premium should be attached to oil imports. But since no such premium is incorporated in the price at which imported oil is sold, the result will be underinvestment in the development of domestic alternatives.

Yet another kind of market imperfection that might justify government support of innovation – in many ways the most important of all – has to do with the problem of nonappropriability. The qualitative idea behind the nonappropriability claim is straightforward enough: private firms will underinvest in R&D if their competitors are able to gain free access to the results of these investments. Others – so called free riders – will not invest at all. (Why pay if you can get it free?) So the system of laws and regulations that protects private rights of ownership of intellectual property is vital for ensuring the flow of private innovation. If this system is perceived to be flawed – if, for example, the courts are viewed as unwilling to enforce the rights of patent holders against patent infringers, or if law enforcement organizations are perceived as being reluctant to prosecute firms or individuals engaged in industrial espionage – private investment in the development of new technical knowledge will be inhibited.

Of course, for certain kinds of R&D society decides as a matter of policy not to protect private intellectual property rights. In particular, for one extremely important category of technical knowledge – the fundamental knowledge about natural phenomena that results from basic scientific research in the Bohr quadrant – patenting is not generally allowed, and there are, in effect, no private intellectual property rights. In this case, society judges that the benefits of making such knowledge freely available to all exceed the benefits of granting private rights to the knowledge and harnessing market forces to develop it. As we have already noted, this type of research is therefore financed mostly by the government, and the results are published in the open literature.

For R&D in the Edison quadrant that leads directly to new or improved commercial products and services, however, the opposite choice is made: Private ownership of the intellectual property is permitted and protected as a matter of law and policy, and private financing is relied on for the most part. Some of what firms learn in the course of their Edisonian R&D is so particular to their own internal routines and capabilities that it is of little interest to would-be poachers. But in other cases the knowledge may be very valuable to rivals, and the protections against unauthorized use that are provided by the patent and copyright system and by the commercial secrecy laws are of great importance in persuading firms that it is worth investing in this kind of R&D.

In theory, there is an alternative: The R&D in the Edison quadrant could be financed with public funds, and the results, instead of being held proprietary, could simply be given away. This would certainly ensure that the new knowledge was more widely disseminated. But for Edisonian R&D the disadvantages of such a scheme would be overwhelming. Even if the necessary public funds were forthcoming (extremely unlikely), the scheme would still suffer from two crippling drawbacks. No firm would have a strong incentive to exploit the results of the R&D, because every other firm would have access to the same information at the same price (i.e., free). Moreover, the public authority would be placed in the position of having to choose which particular technology to develop – a task for which it would be spectacularly unsuited, lacking the sophisticated knowledge of the marketplace that is necessary to make sensible R&D resource allocation decisions. Far better that decisions on what to do be made by private firms, who know much more about the marketplace, and far better that these firms be given strong incentives to invest their own resources in the R&D by allowing them to own the results. Even though diffusion of the knowledge will be restricted, society will be better off as a result.

So for much of the R&D on Stokes' map the choice concerning intellectual property rights is fairly clear. Most of the fundamental research in the Bohr quadrant is placed in the public domain and is financed by the government, whereas R&D in the Edison quadrant intended to develop or improve a specific product or process is almost always paid for by for-profit firms responding to market forces, and the resulting knowledge is held proprietary.

The solutions are less obvious, however, for the important category of commercially relevant fundamental research in the Pasteur quadrant – research that is aimed at strengthening the technological base of entire sectors of the economy. This kind of R&D tends to be longer term – a time horizon of five to ten years is not unusual – and can range from fundamental studies (e.g., research on the properties and behavior of high-temperature superconducting materials) to the development of prototype or demonstration technologies (in our taxonomy, the latter would straddle the Pasteur-Edison boundary). Research in support of technological standards for industry is another example of work in this category. The problem is that neither the Bohr-type government financing model nor the company-financed Edisonian market model is quite right in this case. This kind of research is typically of interest to many firms, so allowing a single firm to hold the results proprietary is not desirable. Yet, if private intellectual property rights are weakened, individual private firms will underinvest because of the free-rider problem. On the other hand, the government is unlikely to be able to allocate funds among the alternative research possibilities wisely, since eventual commercial viability is a key decision criterion, and government officials are typically too distant from the marketplace to be able to judge this effectively. At best, such decisions are likely to be influenced by political considerations; at worst, they will be tainted by logrolling and pork barrel politics. Unless great care is taken, government support can quickly degenerate into subsidies for particular companies.

A variety of ad hoc government and industry-funded approaches have been used to fund this sort of work in the past. Government support was often provided under the umbrella of one or another of the government's statutory missions, especially national security. We have previously stressed the importance to today's information industries of DOD-funded long-term research in fields such as electronics and computer science in the post-war period. Today much knowledge is flowing in the opposite direction, with the military relying more heavily than before on private industry for important technological components and systems. Still, DOD support of so-called "dual-use" technologies that have both military and commercial applications, for example, high-resolution flat-panel displays, is an important force for technological advance in the civilian economy.

In the energy field, the national laboratories of the Department of Energy have long carried out R&D with direct relevance to commercial products and services, often to the point of developing prototypes and demonstration projects. This was usually justified on the grounds of its expected contribution to a public mission – national security or environmental protection. More recently, as we have noted, government technology policy has been focusing more on how to increase the productivity and competitiveness of the civilian economy, and the government has developed a new array of cost-sharing arrangements with firms in a broad range of industries to support commercially-oriented research and development directly. Examples include the Commerce Department's Advanced Technology Program and Manufacturing Extension Partnerships. An important consideration in the design of such programs has been to ensure that the decisions on how to allocate the government's funds are informed by relevant knowledge about marketplace trends, knowledge which is generally only available to private firms who are active in the market.

Consortia between the government and private companies is another approach that has increasingly come into vogue as a way of introducing private sector expertise into R&D planning. Examples of such public/private consortia include the Advanced Battery Consortium, the Partnership for a New Generation of Vehicles, Cooperative Research and Development Agreements (CRADAs) with the DOE national laboratories, and SEMATECH, a DARPA-sponsored consortium to advance semiconductor manufacturing. Some of this commercially-relevant R&D lies unambiguously in the Pasteur quadrant; it is fundamental in nature, and the resulting knowledge is broadly applicable. But other projects are more narrowly applicable to particular products and services – that is, that are Edisonian in nature. Some have criticized the government for venturing too far into what should be the province of the private sector, tilting the playing field in favor of particular firms, and making resource allocation decisions better made by private companies. This is an area in which it is difficult to find clear distinctions and make rules regarding what is in and what is out of bounds for government support. The task is made even more difficult by the changing character of the innovation process in which, as we have seen, the distinction between "front-end" research and "downstream" implementation is itself becoming increasingly blurred.

Policy Instruments for Government Support of Innovation

In the previous sections we examined the relationship of R&D to innovation and discussed the various rationales for government involvement in the innovation process. In this section we review the policy instruments that are available to the government to achieve its goals. Up to now we have emphasized the government's role in funding R&D directly, but it is important to recognize that there are many other ways in which the government can and does influence the innovation process. Some of these policy instruments are general – that is, they are not targeted on a particular industry or technology, but their influence is nonetheless very great. At the most general level, the government's management of the economy has an indirect but profound effect on the pace of innovation. If the general economic environment is perceived to be a significant source of additional investment risk – as occurs, for example, during periods of high inflation or when interest rates or currency exchange rates are very volatile – investors will be correspondingly less likely to make long term investments of all kinds, including investments in innovation. Thus the government's success (or lack of it) in using fiscal and monetary policies to provide a stable macroeconomic environment is one of the most important influences on the innovation process.

A second extremely important general policy instrument is public investment in education. Strong education programs at all levels, enhancing the supply of trained scientists, engineers, and technicians, as well as the general level of education of the workforce, greatly enable innovation. Third, as we have seen, government policies regarding the ownership of intellectual property have a tremendous influence on innovation. Changes in these policies can have a large impact. For example, the United States passed a law in 1980 (the Bayh-Dole Act) that gave universities the right to own patents derived from government-sponsored research carried out in their laboratories. Previously all such intellectual property had belonged to the government. The Bayh-Dole Act has resulted in a greatly increased rate of patenting by universities, and a corresponding increase in the rate of technology licensing – often to startup companies – with the universities and the individual inventors sharing the royalties. Many other countries do not have comparable policies and not surprisingly have a much lower rate of university-originated innovation.

Aside from these very important general policy measures, the government also has many instruments to encourage innovation in a more targeted way. The most obvious is by direct R&D grant or contract. The decisions about what research to undertake within a given field and who should undertake it are often delegated by government agencies to the scientific community itself, through the mechanism of peer review of competitive grant proposals. Performers of this research may be universities, not-for-profit organizations such as hospitals, government laboratories or private firms. This approach is particularly appropriate for fundamental research in the Bohr and Pasteur quadrants. The National Science Foundation and National Institutes of Health allocate much of their research funds in this way. In some cases, especially for research in the Pasteur quadrant conducted in support of one or another of the government's missions, government officials play a more active role in resource allocation decisions, although usually again with the benefit of outside technical advice. Sometimes instead of conducting a competition the government may decide to select a research organization as a "sole source" contractor.

Mechanisms for supporting "pre-competitive" technology development are necessarily somewhat different. These activities are concerned with technology implementation in the commercial sector and, as we have noted previously, must therefore involve private firms in a major way. Federal programs of this type include the Department of Commerce's Advanced Technology Program, DARPA's dual-use technology programs, and DOE's CRADA mechanism for cooperating with industry. In each case, greater private sector involvement in R&D planning is encouraged, and direct government sponsorship is typically augmented by industry cost-sharing. An older and very successful model has been the Agricultural Extension Service run by the Department of Agriculture, which gives valuable assistance to farmers in the process of adopting new technology.

In some cases, the objective of government action goes beyond creating new technology options to demonstrating their practical use in the private sector. In Chapter 12, we discussed the creation of the Synthetic Fuels Corporation as a quasi-government organization, authorized and funded to support the demonstration of new technology at plant scale. In the late 1980s, proposals were put forward for a "Civilian Technology Corporation" to support the development of pre-competitive technologies that would improve the international competitive position of the United States in key technologies such as semiconductor manufacturing equipment.

Indirect R&D Incentives

The government also has available a wide range of indirect measures that serve to encourage private industry to undertake R&D. Tax incentives are one such approach. For many years, Congress has granted industry a tax credit for expenditures on R&D facilities and equipment. And as we have seen in earlier chapters, from time to time Congress has also enacted targeted tax incentives for particular activities or technologies such as gasohol, photovoltaics, and home energy conservation.

A second indirect approach involves the use of regulation. We mentioned previously the regulations that entitled oil produced from domestic enhanced-oil recovery projects to be sold at the world oil price rather than the lower regulated domestic oil price. By far the most significant use of regulation to drive technology development in a particular direction, however, is in the environmental field. The Environmental Protection Agency frequently issues regulations on future emissions that are intended to drive emission control technology. For example, new pollution-control rules applicable to heavy-duty diesel-fueled trucks and buses scheduled to enter into force in 2006 will require engine manufacturers and oil refiners to develop and adopt new technologies in order to comply. The law promulgating standards for gas mileage, the Corporate Automobile Fuel Economy (CAFE) standards, has served as an incentive for automobile manufacturers to push technology that will improve fuel efficiency.

Other indirect incentives have also been used from time to time. These include special government loan guarantees and special purchase programs, for example, for synthetic fuel. Often the government will use its purchasing power to encourage production, for example by purchasing electric autos or photovoltaics for federal buildings. Federal purchases are often seen to be a useful part of a buy-down effort to help drive down the cost of production of new technology. In Chapter 13, on fuel cells, we discussed some of the difficulties involved in deciding whether such buy-down programs are likely to be successful.

How should the government choose among these various policy instruments? There are no ironclad rules. In general, however, direct funding support makes sense for fundamental research in the Bohr and Pasteur quadrants and for R&D directed to federal uses. The indirect mechanisms make progressively more sense as the R&D activity moves toward encouraging pre-competitive technologies and the demonstration of the technology in the marketplace. The closer the technology is to commercialization,

Billions of constant 1996 PPP dollars

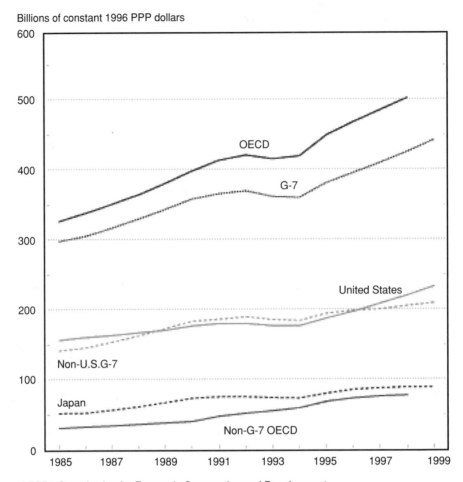

OECD - Organization for Economic Cooperation and Development.

PPP - purchasing power parity.

NOTE: Non-United States G-7 countries are Canada, France, Germany, Italy, and the United Kingdom.

Source: Reprinted from National Science Board, *Science and Engineering Indicators – 2002*.

Figure 15.4. U.S., G-7, and OECD countries' R&D expenditures.

the less the government should be involved in making decisions about which specific technologies to support.

International Comparisons

The government's role in the innovation process varies from one country to another. We conclude this chapter with a few comparisons between the United States and other developed countries, especially Japan and other members of the so-called G-7 group of advanced economies.

The most important point to appreciate is that there are significant differences in how governments approach the task of encouraging innovation in their economies. In Japan,

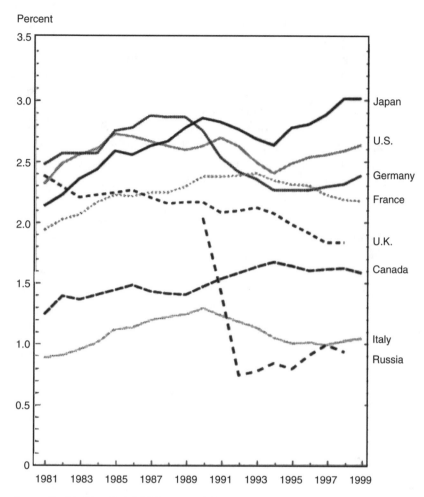

Percent

Source: Reprinted from National Science Board, *Science and Engineering Indicators – 2002*.

Figure 15.5. International comparisons of R&D as a percentage of GDP.

for example, there has historically been a very close relationship between government and industry – so close, in fact, that it has often been difficult to differentiate between government policy as set forth by agencies, notably the Ministry of Industry and Trade and Industry (now renamed the Ministry of Economy, Trade, and Industry, or METI), and the strategies of large companies. The same has also largely been true of some other Asian economies.

On the other hand, in many European countries (with the notable exception of France), governments have traditionally maintained more of a hands-off role. European integration has meant that the European Commission, headquartered in Brussels, has an important voice in many aspects of innovation policy. This adds a heavy layer of bureaucracy to the decisionmaking process that is unlikely to foster agility.

Most experts believe that Europe is currently well behind the United States in its capacity for innovation. Three reasons are commonly cited for this lag. The first is that Europe, which has lacked an extensive venture capital market and has a weaker record of university-industry cooperation, has a much less well-developed tradition of new technology-based enterprise formation. Several countries, notably Germany and the United Kingdom, have moved to correct this. Second, European industry has not adopted information technology as extensively or as rapidly as has U.S. industry. This has implications both for the efficiency of operations and for the ability to implement the parallel approach to innovation described earlier. Finally, as indicated in Figures 15.4 and 15.5, Europe and Japan spend less on R&D than the United States in absolute terms (although not, in the case of Japan, as a percent of the GDP). Spending more or less is not the ultimate test, of course; the real test is whether the R&D is effective in contributing to innovation in the economy. Here the U.S. experience (both in terms of research and education) is regarded, for all its limitations, as being the best in the world.

This brief comparative assessment of the state of innovation should not induce complacency in the United States. The leadership of the United States in the 1990s was not preordained, nor should it be assumed to be permanent. As previously noted, in the late 1980s Japan was viewed as being well ahead of the United States and Europe in its capacity to innovate in manufacturing industries like autos and electronics. Today, European leaders are aware of the U.S. position, and they are taking aggressive action to redress the balance.

16

Conclusions

In the academic world, technology-oriented students mainly learn in "stove pipes" defined by established disciplines in the natural and social sciences and in engineering. This compartmentalization is understandable because the foundations of the disciplines are most easily learned separately and best taught by specialists. One must learn calculus, chemical kinetics, microeconomics, or system control before being able to address the complex relationships between these subjects that may arise in practical applications of technology. The result, however, is that students learn about the pieces of a problem rather than the whole. Moreover the learning is typically a solitary activity, and rarely depends upon the cooperation of a group.

Yet real problems are an inseparable mix of technological, economic, environmental, and political factors. As we have seen, successful application of technology frequently requires the synthesis of all these considerations. When this synthesis is absent or is not credible, the technology will fail to live up to its potential and may fail completely. And successful syntheses require individuals to work together to bring different skills to bear on the problem. We teach individual disciplines, but the resolution of most real problems requires that different disciplines be jointly brought to bear. It is as if people were taught how to play the individual musical instruments of an orchestra and then expected immediately to perform a symphony.

Students understand that they will face these sorts of problems in the course of their professional careers. We have found that they are eager to be exposed to these complexities and enjoy learning about the "tools" that will help them succeed in applying technology to meet society's needs, whether as entrepreneurs or as public officials. But because the conventional educational experience stresses individual effort to achieve command of a discipline, the students are often reluctant to work with or depend upon others trained in different disciplines. Given the opportunity to work together in an interdisciplinary group, however, the reluctance quickly turns to mutual respect as the power of a team to address a complex issue becomes manifest.

The first purpose of this book, therefore, has been to introduce interested students in the natural sciences, engineering, and the social sciences to the complexity of technology application problems. In each of our case studies, we have considered systematically the technical, economic, environmental, and political issues that arise. The cases have ranged from examining the causes of technical failure (e.g., nuclear energy) to understanding how regulatory constraints can shape the development of new technology (e.g., global warming). We believe that anyone who has read this book will not easily ignore these connections in the future.

A second purpose of the book has been to convince the reader of the value of explicit analysis in addressing complex problems of technology application. In the first chapter, the problem of "paper versus plastic" illustrated the confusion that often pervades public debates about technology (in that case the confusion stemmed from differences in the way the system under consideration was defined). We believe that explicit analysis can help to reduce the risk that public resources or private capital will be wasted in the course of setting or pursuing a goal. Objective analysis, in which the benefits and costs in both private and public terms are explicitly estimated, helps point the way to successful and socially desirable uses of technology.

Of course, specifying the benefits and costs does not automatically lead to the successful resolution of a problem. As we have seen, in cases such as acid rain or the siting of a nuclear waste repository the benefits and costs are perceived and experienced very differently by different groups. Indeed, most of the issues we have addressed in this book have this property (global warming being perhaps the most prominent example). In such cases it is inevitable that the solution, if and when it comes, will be worked out in the political arena. And the sooner that scientists and engineers appreciate the inevitable role of politics in technology applications, the more successful they are likely to be. Explicit analysis is essential to inform both public and private decisions, but it is by no means the whole story (recall the case of gasohol presented in Chapter 2).

Many people disdain analysis, but do so for different reasons. Some, often politicians or business people (and the lawyers who serve them), believe it most important to work for the interests of the people they represent – voters or stockholders. They use analysis to argue for a particular position. Others, usually environmentalists and members of public interest groups, believe that analysis is invariably used to defend the status quo and does not adequately address the long-term public interest, distributional issues, and externalities. We encountered this view in Chapters 2 and 3, where tax credits were adopted in order to achieve the goal of encouraging biomass and solar energy, despite an unfavorable cost/benefit analysis. But good intentions are not enough to justify technology investments, public or private. If they were, there would be no way to exclude alternatives that have popular or political support. Under such circumstances, scarce capital would likely be allocated to projects that are less efficient, in the sense of maximizing public or private return. In the private sector, investors demand an adequate return on invested capital, so there is generally an expectation that some sort of economic analysis (such as the discounted cash flow analysis discussed in Chapters 3 and 4) will be presented to justify an investment. But in the public arena, "other people's money" is at stake, and frequently no objective analysis is offered. There is nothing more frustrating to experience than public debates between parties with different interests who do not have and are not willing to adopt a common analytic framework. Without such a framework, resolution is virtually impossible, and the public policy process becomes inchoate.

Our strong defense of explicit and, wherever possible, quantitative analysis does not mean that we oppose advocacy or normative judgments about who should gain and who should pay for particular initiatives – for example, in setting environmental regulations or in shaping government programs that encourage technology through

R&D support or subsidy. Rather, we believe that this advocacy will be both more responsible and more effective if it is based on objective analysis. There may be no more effective demonstration of the value of analysis in structuring important debates than the accounting identity offered by Yoichi Kaya to help clarify the expectations that can logically be held simultaneously about future economic growth, energy intensity, and greenhouse gas emissions (see Chapter 6).

In view of the importance of analysis, a related purpose of this book has been to introduce the reader to particular tools that are helpful in analyzing different technology issues. Our intention has not been to develop rigorous analytical methodologies, but rather to illustrate how particular tools can be useful and in some instances are essential for understanding an application of technology. Examples include:

Chapter 1: Defining the system (paper versus plastic cups)
Chapter 2: Energy and mass balances (gasohol)
Chapter 3: Present value analysis (wind, solar hot water, and photovoltaics)
Chapter 3: Sensitivity analysis (wind)
Chapter 4: Levelized lifecycle costs (coal-fired electricity generation)
Chapter 5: Cost/benefit analysis (acid rain abatement)
Chapter 6: Accounting identities for relationships among variables (global warming)
Chapter 10: Economic analysis of exhaustible resources (natural gas)
Chapter 11: Probabilistic risk assessment (LNG)
Chapter 13: Learning curves (fuel cells)
Chapter 14: Linear regression models (energy forecasting)

Our description of these analytical tools should, at the very least, inform the reader about techniques that are useful in the analysis of technology applications. More importantly, we hope that exposure to these tools will encourage the reader to seek expert advice when a specific technique seems pertinent to a particular problem he or she has encountered.

The wide range of case studies presented in this book demonstrates that there is no uniform template that can be used to analyze all technology applications. Nevertheless, certain elements of analysis are essential in almost every situation. We summarize them here:

1. *Scope the problem*: Quantitative estimates should be developed for all of the key outputs (benefits) and inputs (costs) associated with the application. This exercise should include both direct and indirect effects. Direct effects are typically reflected in the prices paid or received by the technology developer or owner. Indirect effects (environmental and social) are external to commercial transactions and include effects borne by those who are not parties to these transactions.

2. *Create options*: Realistic alternative courses of action should be developed and compared with the course of action under consideration. Analyzing alternatives is the best safeguard against adopting a path that entails a less-than-optimal use of public or private resources.

3. *Address uncertainties explicitly*: The implications of uncertainty should be analyzed by explicitly considering a range of outcomes, or "scenarios," with probabilities attached to each. This process will give an indication of the range of possible gains and losses associated with any given course of action.

4. *Reflect on what is left out*: Analysis cannot encompass all reality. The best claim for a well-conducted analysis is that it focuses attention on key assumptions and the essential variables that govern system choice. But analysis is difficult to extend to certain critical aspects of reality, most importantly political and social attitudes and consequences. To avoid missing the forest for the trees, the responsible analyst and wise decisionmaker is well advised to ponder those aspects of the situation that lie outside the analytical framework that he or she has adopted.

A fourth purpose of this book stems from our conviction that decisions about technology applications will gain effectiveness and credibility if technologists play an active role in the decision-making process. A common view is that technical people – engineers or scientists – are unlikely to lead in shaping public debates or in resolving technology application issues; the expectation is that the leadership roles will be played by those with a law or business background. We believe that this attitude, to the extent that it is justifiable and not mere prejudice, is attributable to the relatively narrow perspective with which technical people tend to approach problems. Technically trained individuals typically prefer to stick to what they know and are usually reluctant to speak out on or make assumptions about aspects of a problem that they have not themselves studied in depth. While professional caution is both admirable and essential, we believe that it too often degenerates into a parochial, insular approach that unduly constrains the contribution that technical people are able to make.

Accordingly, this book is intended to encourage scientists and engineers to be more ambitious about the roles they play in major technology-related issues. To do this effectively, the issue must be seen in the broadest context, and the technologist must be willing to work with others skilled in finance, law, the environment, management, politics, and other specialties in order to achieve progress. The examples in this book have been taken from the field of energy and the environment, but the same argument can be made for other fields such as biology and medicine, information and communications, food and agriculture, and national defense. In all such fields, we believe that when technologists gain an appreciation for the complexities of applying technology in the real world, and when they learn to work with others to integrate methods, perspectives, and attitudes different from their own into the technology development process, they will have taken a major step towards realizing the potential of technology to benefit humanity, as well as their own potential as effective professionals and contributors to society.

Index